Polymerase Chain Reaction Research

Edited by Murat Aycan and Mustafa Yildiz

Published in London, United Kingdom

Polymerase Chain Reaction Research
http://dx.doi.org/10.5772/intechopen.106100
Edited by Murat Aycan and Mustafa Yildiz

Contributors
Abebe Abuhay Zemene, Ali Dino, Ayaz Ahmed, Jacqueline Murphy, Jin Cai, Jingwen Liu, Jinhong Chen, Jishun Yang, Kate Herr, Maliha Wajeeh, Mingke Wang, Nadia Jamil, Saifullah Khan, Shezaib Siddiqui, Venkata Vepachedu, Waqar Hassan, Xiaoyu Geng, Xuelu Yu, Zohaib Khan

First published in London, United Kingdom, 2024 by IntechOpen
IntechOpen is the global imprint of INTECHOPEN LIMITED, registered in England and Wales, registration number: 11086078, 167-169 Great Portland Street, London, W1W 5PF, United Kingdom

British Library Cataloguing-in-Publication Data
A catalogue record for this book is available from the British Library

Additional hard and PDF copies can be obtained from orders@intechopen.com

Polymerase Chain Reaction Research
Edited by Murat Aycan and Mustafa Yildiz
p. cm.

This title is part of the Biochemistry Book Series, Volume 55
Topic: Chemical Biology
Series Editor: Andrei Surguchov
Topic Editors: Murat Şentürk and Deniz Ekinci

Print ISBN 978-1-83768-900-2
Online ISBN 978-1-83768-901-9
eBook (PDF) ISBN 978-1-83768-902-6
ISSN 2632-0983

For EU product safety concerns:
IN TECH d.o.o., Prolaz Marije Krucifikse Kozulić 3, 51000 Rijeka, Croatia, info@intechopen.com or visit our website at intechopen.com.

IntechOpen Book Series
Biochemistry
Volume 55

Aims and Scope of the Series

Biochemistry, the study of chemical transformations occurring within living organisms, impacts all of the life sciences, from molecular crystallography and genetics, to ecology, medicine and population biology. Biochemistry studies macromolecules - proteins, nucleic acids, carbohydrates and lipids –their building blocks, structures, functions and interactions. Much of biochemistry is devoted to enzymes, proteins that catalyze chemical reactions, enzyme structures, mechanisms of action and their roles within cells. Biochemistry also studies small signaling molecules, coenzymes, inhibitors, vitamins and hormones, which play roles in the life process. Biochemical experimentation, besides coopting the methods of classical chemistry, e.g., chromatography, adopted new techniques, e.g., X-ray diffraction, electron microscopy, NMR, radioisotopes, and developed sophisticated microbial genetic tools, e.g., auxotroph mutants and their revertants, fermentation, etc. More recently, biochemistry embraced the 'big data' omics systems.
Initial biochemical studies have been exclusively analytic: dissecting, purifying and examining individual components of a biological system; in exemplary words of Efraim Racker, (1913 –1991) "Don't waste clean thinking on dirty enzymes."
Today, however, biochemistry is becoming more agglomerative and comprehensive, setting out to integrate and describe fully a particular biological system. The 'big data' metabolomics can define the complement of small molecules, e.g., in a soil or biofilm sample; proteomics can distinguish all the proteins comprising e.g., serum; metagenomics can identify all the genes in a complex environment e.g., the bovine rumen.
This Biochemistry Series will address both the current research on biomolecules, and the emerging trends with great promise.

Meet the Series Editor

Andrei Surguchov, Ph.D., joined Baylor College of Medicine, Houston, TX, as a faculty member in 1992, where he studied the mechanisms of the genetic control of lipid metabolism. At the University of Utah, his research interests focused on cloning new genes encoding retinal proteins. He studied molecular and cellular mechanisms of neurodegenerative diseases and retinal degeneration at Washington University, St. Louis. Currently, his research focuses on the structure-function relationship of proteins involved in neurodegeneration and ocular diseases. Andrei Surguchov is an Editor-in-Chief at Biochemistry Research International and Associate Editor in several biomedical journals.

Meet the Volume Editors

Dr. Murat Aycan is a Japan Society for the Promotion of Science (JSPS) postdoctoral fellow working in the Faculty of Agriculture, Niigata University, Japan. He received his BSc in Biology from Sakarya University, Turkey in 2012, an MSc in Agronomy from Ankara University, Turkey in 2015, and an MEXT scholarship to visit Niigata University during his doctoral studies. Dr. Aycan received a Ph.D. in Agronomy from Ankara University in 2018 for developing two salt-tolerant bread wheat cultivars. He also discovered that extra-nuclear transgenerational inheritance gives offspring better salinity tolerance from the maternal parent. Since 2018, Dr. Aycan has been an assistant professor and JSPS postdoctoral scholar at the same university, continuing his research. His study focuses on salinity and high-temperature stress tolerance molecular mechanisms, plant breeding for tolerance, transgenerational environmental effects, extra-nuclear inheritance, and plant stress memory.

Mustafa Yildiz obtained his MSc in Agricultural Sciences from Ankara University, Turkey in 1996 with a thesis on the effect of cell structure on yield and sucrose concentration in sugar beet. He trained in plant biotechnology at Osaka Prefecture University, Japan in 1998. He received his Ph.D. from Ankara University in 2000 with a thesis on shoot regeneration and *Agrobacterium tumefaciens*-mediated gene transfer in flax. He is currently a professor in the Faculty of Agriculture, Department of Field Crops, Ankara University. Dr. Yildiz's research interests include plant tissue culture, plant biotechnology, molecular markers, *Agrobacterium tumefaciens*-mediated gene transfer, plant stress physiology, plant immune systems, plant defense mechanisms, and plant breeding. He has more than seventy scientific publications, two books, and nine book chapters to his credit. He was awarded first place in the International Sunflower Project Market by the International Sunflower Association for his project titled "A New Environmental Friendly Production Method in Sunflower for High Seed and Crude Oil Yields."

Contents

Preface

The polymerase chain reaction (PCR) is a groundbreaking molecular biology technology that has changed the face of biological study and diagnosis. This book is a comprehensive guide through the principles, applications, and breakthroughs in PCR.

Kary B. Mullis developed PCR in the mid-1980s, and it has since become a vital tool in laboratories worldwide. Its ability to massively increase DNA has transformed genetics research while paving the path for advances in health, forensics, and biotechnology. This book examines PCR's essential concepts and recent developments.

The chapters are designed to explain PCR methodology, catering to both the novice seeking a firm basis and the seasoned researcher exploring advanced applications. This book explores the PCR's multidisciplinary character, revealing its applications in domains such as genetics, clinical diagnostics, environmental research, and more. The narrative is enhanced with real-world examples and case studies demonstrating PCR's impact on our understanding of the biology sciences.

It is critical to stay current on the newest advancements in the rapidly changing field of molecular biology. As a result, this book provides an introduction and digs into developing trends and future prospects in PCR research. With contributions from industry professionals, you'll gain insights into latest technology and new applications, ensuring you stay ahead of the curve in PCR breakthroughs.

This book is a useful resource for students, researchers, and professionals in the biological sciences, sparking curiosity and encouraging a better understanding of the extraordinary PCR process.

Happy reading, and may your experiments be as fruitful as your curiosity is.

Murat Aycan, Ph.D.
Faculty of Agriculture,
Laboratory of Biochemistry,
Niigata University,
Niigata, Japan

Mustafa Yildiz
Faculty of Agriculture,
Department of Field Crops,
Ankara University,
Ankara, Turkiye

Section 1

Introduction

Chapter 1

Introductory Chapter: A Molecular Photocopying Polymerase Chain Reaction (PCR)

Murat Aycan and Mustafa Yildiz

1. Introduction

Following its discovery in 1983 by American scientist Mullis, the polymerase chain reaction (PCR) has enabled the reproduction of particular DNA sequences *in Vitro* [1]. PCR, also known as polymerase chain reaction or molecular amplification, has utility in diverse fields and contexts. The discovery of the PCR brought about a significant transformation in the field of biological science. For the first time, it facilitated the precise identification and synthesis of substantial quantities of DNA. The utilization of PCR-based methodologies has significantly advanced substantial scientific undertakings, exemplified by the Human Genome Project [2].

PCR is an enzyme technique that efficiently amplifies a targeted DNA fragment from a diverse mixture of DNA molecules. In order to conduct a PCR assay, it is necessary to have template DNA, primers, nucleotides, and DNA polymerase. The DNA polymerase is an essential enzyme responsible for catalyzing the formation of the PCR product by connecting individual nucleotides. Nucleotides are present throughout DNA and encompass four distinct bases, namely adenine, thymine, cytosine, and guanine (A, T, C, G). The aforementioned components serve as the fundamental units utilized by DNA polymerase in the synthesis of the final product obtained by PCR. The primers utilized in the reaction are responsible for specifying the precise DNA product that is to be amplified. The primers utilized in this study are brief DNA fragments, typically consisting of approximately 20 base pairs. These primers possess a specific sequence that is complementary to the target DNA, enabling its identification and subsequent amplification. These structures function as sites of elongation for DNA polymerase throughout the process of DNA synthesis [2].

In order to do DNA amplification, a series of procedural steps must be undertaken. The initial denaturation phase commences at a temperature of 94°C. The denaturation of matrix DNA, which functions as a template for replication, occurs at temperatures over 80°C due to the inability to sustain hydrogen bonding. The second step involves the process of hybridization. The primers exhibit hybridization in a temperature range of 40–70°C. A decrease in temperature facilitates the regeneration of hydrogen bonds, allowing the strands to complement one other and undergo hybridization. Short single-strand sequences that match regions adjacent to the DNA to be amplified exhibit greater ease of amplification compared to long-strand matrix DNA. Elevated temperatures during hybridization processes result in enhanced selectivity and specificity of hybridization. The temperature required for the third step, which involves the synthesis of the complementary strand, is 72°C. At a temperature of 72°C, Taq polymerase utilizes deoxyribonucleoside triphosphates present in the reaction mixture to facilitate the replication of primed single-stranded DNAs. The process of selectively synthesizing template DNA downstream of primers takes place. The fragments that were synthesized in the preceding cycle are incorporated into the matrix during the subsequent cycle. After several cycles, the prevailing species aligns with the DNA sequence located between the primers. The process of generating 0.1 μg of DNA for examination typically requires 20–40 cycles, with each cycle theoretically doubling the DNA amount.

The utilization of DNA copies generated via PCR amplification finds extensive application in various domains of medical and plant research. Furthermore, this technology can also be utilized for the identification and detection of infectious diseases, as well as for a diverse range of research endeavors in the field of molecular genetics. One of the most remarkable applications of PCR. This technique enables the isolation and purification of genes without resorting to conventional molecular cloning methodologies, which involve the insertion of a DNA library into a plasmid vector and subsequent transformation of a bacterial strain. The clones undergo additional screening. PCR enables a more rapid and controlled realization process, characterized by less randomness. As stated in the previous section, the extraction of messenger RNAs (mRNAs) has the potential to facilitate the replication of complementary DNAs (cDNAs). The procedure involves the synthesis of DNA from RNA by retrovirus reverse transcriptase. Initially, the extraction of total RNAs is performed. Messenger RNAs (mRNAs) are isolated from total RNA using affinity chromatography with oligodT (polyT oligonucleotide) due to the presence of a 3′polyA sequence in mRNAs. Subsequently, reverse transcriptase facilitates the synthesis of complementary DNA (cDNA) molecules from individual messenger RNA (mRNA) strands. mRNAs are subject to hydrolysis by alkaline conditions, RNase enzymes, or elevated temperatures subsequent to reverse transcription. The following procedures are implemented within the thermal cycler enclosure. In the initial temperature cycle, the replication of single-stranded cDNAs is facilitated by DNA polymerase [3, 4].

Quantitative polymerase chain reaction (qPCR), which was pioneered in the mid-1980s, is a technique used to quantify the quantities of DNA or RNA in biological specimens. The methodology employed in this study involves the detection of a fluorescent signal that is directly proportional to the amplification of the PCR result with each subsequent cycle. The experimental setup requires the utilization of a temperature cycler and an optical apparatus designed for fluorescence measurements. The process of synthesizing a nucleotide probe enables it to specifically hybridize with the specific DNA of interest located between the primers. The predominant method employed is reverse transcription polymerase chain reaction (RT-PCR). Quantitative PCR is utilized to determine the precise quantity of RNA or DNA present. RNA or DNA standards are utilized to assess the relative quantities in semi-quantitative or competitive PCR methodologies. The predominant method employed is RT-PCR. Standards can be categorized as either internal or external, with external standards further classified as homologous or heterologous. The established protocol involves the utilization of RNA (or DNA) in the extracted RNA or provided in predetermined quantities to the reaction mixture. Simultaneous amplification of both standard and target RNA is conducted. Therefore, there is a competition between the amplification of conventional DNA and the DNA of interest. As the standard quantity increases, the amplification of the RNA of interest will decrease, resulting in a smaller final product. The process for analyzing PCR samples should possess the capability to differentiate between the standard and the RNA of interest, as well as to assess the relative quantity of the DNA of interest in comparison to the standard [5].

PCR is a highly effective technique for diagnostic purposes. The utilization of this technique is prevalent in the identification of genetic diseases. The process of gene amplification enables the identification of deleterious mutations associated with a genetic disorder, providing valuable information regarding their specific locations, sizes, and characteristics. Electrophoresis of PCR products or the integration of PCR with other methodologies might facilitate the identification of deletions, inversions, insertions, and point mutations. Similar to acquired immunodeficiency syndrome (AIDS), hepatitis C, and chlamydia, the PCR technique has the capability to identify and diagnose several infectious diseases caused by viruses, bacteria, parasites, and other pathogens. While other diagnostic methods have the capability to identify these illnesses, PCR is particularly advantageous due to its ability to generate highly accurate and rapid results using minuscule biological samples in which the presence of the pathogen is typically not evident [6].

Genetic illnesses encompass the identification of gene mutations. There are several issues that arise. Examples of basic types of errors in data manipulation include insertions and deletions. The mutation results in alterations to either the size of the gene or specific segments within the gene. A comprehensive understanding and characterization of the mutation is sufficient for the purpose of gene amplification. In the context of insertions, it has been observed that the DNA PCR products derived from patient samples exhibit greater length compared to those derived from healthy individuals. The act of deleting something yields outcomes that are contrary to the original intent or desired outcome. The diagnosis can be determined by analyzing the electrophoretic patterns of PCR products and their corresponding sizes. Identifying inversions and point mutations presents greater difficulty. There is minimal disparity in size between healthy and diseased DNA whether considering an inversion or a point mutation. Therefore, it is not possible to maintain the condition of the PCR product size in order to get the desired result. Therefore, it is necessary to employ PCR-complementary techniques. Three potential approaches for genetic analysis include Southern blot, restriction fragment length polymorphism (RFLP), and mismatch detection. In order to detect the mutation in the southern blot, a hybridization process is performed using a radioactive isotope or a fluorochrome-marked oligonucleotide probe, which is applied to the PCR result. The effectiveness of this method has been demonstrated in the context of inversions, as evidenced by studies conducted by Li et al. [7] and Martell et al. [8].

The presence of viral or microbial contamination, such as bacteria or parasites, invariably results in the deposition of genetic material into the host organism. The PCR method exhibits a higher degree of efficacy in the identification of pathogens inside biological samples, mostly attributed to its exceptional levels of sensitivity and specificity. The process of PCR diagnosis relies on the utilization of primers that are designed to selectively amplify a targeted DNA sequence of a virus or microorganism [9, 10].

The utilization of PCR in genetic fingerprinting exhibits a remarkable level of precision in the identification of species, cultivars, and individuals. This application leverages information pertaining to the structure of genomes. The objective is to amplify nucleotide sequences that are distinctive to each species. Eukaryotes possess a substantial number of these sequences, which provide a wide range of options for accurate and specific identification. In contrast to prokaryotes, eukaryotic genomes consist of both coding and noncoding elements. The process of protein synthesis involves the translation of genetic information encoded in coding sequences. Up to

98% of the genomic DNA in eukaryotes is comprised of untranslated noncoding regions. Individuals belonging to the same species exhibit strikingly comparable coding sequences. Genes confer heritable features and qualities upon a species. Allelic variations give rise to phenotypic variations among individuals, and alleles of a particular gene exhibit sequence variations on the scale of around 1 base pair per 1000 [11, 12]. This chapter presents a compilation of fundamental material pertaining to PCR techniques, with a focus on PCR-based basic concepts, intended for the readers' understanding. PCR is the fundamental analytical technique employed in genetic laboratories, including several fields such as medicine and plant science.

Author details

Murat Aycan[1*] and Mustafa Yildiz[2]

1 Faculty of Agriculture, Laboratory of Biochemistry, Niigata University, Niigata, Japan

2 Faculty of Agriculture, Department of Field Crops, Ankara University, Ankara, Turkiye

*Address all correspondence to: murataycan@agr.niigata-u.ac.jp

References

[1] Mullis K, Faloona F, Scharf S, et al. Specific enzymatic amplification of DNA in vitro: The polymerase chain reaction. Cold Spring Harbor Symposia on Quantitative Biology. 1986;**51**:263-273. DOI: 10.1101/sqb.1986.051.01.032

[2] Garibyan L, Avashia N. Research techniques made simple: Polymerase chain reaction (PCR). The Journal of Investigative Dermatology. 2013;**133**:e6. DOI: 10.1038/jid.2013.1

[3] Freeman WM, Walker SJ, Vrana KE. Quantitative RT-PCR: Pitfalls and potential. BioTechniques. 1999;**26**:112-122. DOI: 10.2144/99261rv01

[4] Joyce C, Quantitative RT-PCR. A review of current methodologies. Methods in Molecular Biology. 2002;**193**:83-92. DOI: 10.1385/1-59259-283-x:083

[5] Schmittgen TD, Livak KJ. Analyzing real-time PCR data by the comparative CT method. Nature Protocols. 2008;**3**:1101. DOI: 10.1038/nprot.2008.73

[6] Somura H, Iizuka N, Tamesa T, Sakamoto K, Hamaguchi T, Tsunedomi R, et al. A three-gene predictor for early intrahepatic recurrence of hepatocellular carcinoma after curative hepatectomy. Oncology Reports. 2008;**19**:489-495. DOI: 10.3892/or.19.2.489

[7] Li MH, Che TC, Kuo TT, Tseng CC, Tseng CP. Real-time PCR for quantitative detection of toxoplasma gondii. Journal of Clinical Microbiology. 2000;**38**:4121-4125. DOI: 10.1128/jcm.38.11.4121-4125.2000

[8] Martell M, Gómez J, Esteban JI, Sauleda S, Quer J, Cabot B, et al. High-throughput real-time reverse transcription-PCR quantitation of hepatitis C virus RNA. Journal of Clinical Microbiology. 1999;**37**:327-332. DOI: 10.1128/jcm.37.2.327-332.1999

[9] Chen W, Martinez G, Mulchandani A. Molecular beacons: A real-time polymerase chain reaction assay for detecting salmonella. Analytical Biochemistry. 2000;**280**:166-172. DOI: 10.1006/abio.2000.4518

[10] Fortin NY, Mulchandani A, Chen W. Use of real-time polymerase chain reaction and molecular beacons for the detection of Escherichia coli O157:H7. Analytical Biochemistry. 2001;**289**:281-288. DOI: 10.1006/abio.2000.4935

[11] Gibson UEM, Heid CA, Williams PM. A novel method for real time quantitative RT-PCR. Genome Research. 1996;**6**:995-1001. DOI: 10.1101/gr.6.10.995

[12] Giesendorf BAJ, Vet JAM, Tyagi S, Mensink EJMG, Trijbels FJM, Blom HJ. Molecular beacons: A new approach for semiautomated mutation analysis. Clinical Chemistry. 1998;**44**:482-486. DOI: 10.1093/clinchem/44.3.482

Section 2

Agricultural Applications

Chapter 2

Detection of Genetic Modification in Crops by Polymerase Chain Reaction

Nadia Jamil, Waqar Hassan Naqvi, Shezaib Siddiqui, Ayaz Ahmed, Ali Dino, Maliha Wajeeh, Zohaib Khan and Saifullah Khan

Abstract

Human beings have a long history of changing crops' characteristics by selective and crossbreeding. However, with the advancement in science and technology, genetic modification in crops is possible to add cross specie characteristics like pest/herbicide resistance, increased yield, etc. Although there are several benefits of genetically modified (GM) crops, genetically modified crops have not gained universal acceptance due to cross specie genes. Therefore, it is mandatory to check the genotype of a crop before import or export and to declare if it is genetically modified or not. This can be done by polymerase chain reaction (PCR), a gold standard for this testing.

Keywords: GMO, PCR, screening, terminator, promoter, event

1. Introduction

Genetically modified (GM) crops, also known as genetically engineered (GE) or biotech crops, are agricultural plants that have been genetically modified through biotechnology tools like oligonucleotide-directed mutagenesis, zinc-finger nucleases, mega-nucleases, transcription activator-like effector nucleases (TALENs), and clustered regularly interspaced short palindromic repeat (CRISPR) systems to enhance desirable traits, such as resistance to pests, diseases, or herbicides, improved nutritional value, or increased yield. These traits are created by inserting specific genes from one organism (same or cross specie) into the DNA of plant [1, 2].

The first genetically modified crop, a tomato that was modified to have a longer shelf life, was approved for commercial use in the United States in 1994. Since then, the adoption of GM crops has steadily increased, with an estimated 190 million hectares of GM crops being grown globally in 2020. The most commonly grown GM crops include soybeans, corn, cotton, canola, and alfalfa [3–5].

The use of GM crops has generated both excitement and controversy. Supporters argue that GM crops can help address global food security challenges by improving crop yields and reducing the use of harmful pesticides and herbicides. Opponents, on the other hand, argue that GM crops could have negative impacts on biodiversity, create new pests or diseases, or crossbreed with wild relatives, potentially leading to unintended consequences. They also raise concerns about the potential risks to human health, as well as the potential for the technology to be used in ways that benefit large corporations at the expense of small farmers and communities.

Overall, the use of GM crops is a complex and highly debated issue that requires careful consideration of both their potential benefits and risks. As such, ongoing research and regulation are necessary to ensure their safe and responsible use.

1.1 Regulation on GMO labeling and GMO detection

Genetically modified organisms (GMOs) have been the subject of regulatory scrutiny around the world due to concerns about their potential impacts on human health, environment, and agriculture. One key aspect of GMO regulation is the labeling of GMO-containing products, which allow consumers to make informed choices about the food they consume. Additionally, GMO detection methods are essential for enforcing GMO labeling requirements, ensuring regulatory compliance, and maintaining consumer confidence.

1.1.1 Key regulatory frameworks in different regions

1.1.1.1 Europe

In the European Union (EU), GMO labeling is mandatory for all food and feed products containing or consisting of GMOs, as well as products produced from GMOs. The EU has established comprehensive regulations for GMO labeling, including Regulation (EC) No. 1829/2003 and Regulation (EC) No. 1830/2003. These regulations require that GMOs and their derived products be labeled with the words "genetically modified" or "produced from genetically modified [name of organism]." Additionally, the regulations specify the threshold levels for GMO presence in food and feed products, below which labeling is not required but GMO presence must be detectable by analytical methods. Polymerase chain reaction (PCR)-based methods are commonly used for GMO detection in the EU, and specific primers have been developed and validated for various GMO events [1, 6, 7].

1.1.1.2 Non-EU countries

Norway, Switzerland, and Russia do not see cultivation or import of GMO crops [2].

1.1.1.3 United States and Canada

In the United States, GMO labeling is not mandatory at the federal level, but there are regulations related to voluntary GMO labeling, while Canada labels it as plant with novel traits, instead of labeling it as GMO. The U.S. Department of Agriculture (USDA) has established the National Bioengineered Food Disclosure Standard, which requires that certain food products containing GMOs be labeled with a disclosure, such as a text statement, symbol, or electronic link. However, this standard has certain exemptions and allows for alternative methods of disclosure, such as QR codes or toll-free telephone numbers, which have been a source of controversy and criticism. Nevertheless, PCR-based methods are commonly used for GMO detection in the U.S., and several GMO events have been approved and validated for detection in food and feed products [8].

1.1.1.4 Asia

Various countries in Asia have also established their own regulations on GMO labeling and detection. For example, Japan has mandatory GMO labeling requirements for certain GM crops, which means it is regulated for GM products, PCR-based methods are commonly used for GMO detection, the same is with Australia, Philippines, and India. China, Pakistan, Indonesia, Singapore, Malaysia, Bangladesh, Vietnam, Thailand, Taiwan, and South Korea have regulations on GMO labeling and detection for imported GM products, and PCR-based methods are widely used for GMO detection in both imported and domestic products. Other countries like Myanmar, Laos, Cambodia, Papua New Guinea, East Timor, Sri Lanka, North Korea, Nepal, and Bhutan are not yet regulated for GM products [9, 10].

Genetically modified organism labeling and GMO detection play crucial roles in ensuring transparency, regulatory compliance, and consumer confidence in relation to GMO-containing products. Legislation on GMO labeling and detection varies across different regions, but PCR-based methods are widely used for GMO detection due to their specificity, sensitivity, and versatility. As GMO technology continues to evolve and new GMO events are developed, it is important for regulatory frameworks to keep pace with these advancements and ensure that GMO labeling and detection regulations remain robust and effective in addressing the concerns and demands of consumers, farmers, and other stakeholders.

Despite the controversy, the use of GM crops continues to grow, with millions of hectares of GM crops being planted worldwide each year. As the technology continues to evolve and new genetic engineering techniques are developed, the debate over the benefits and risks of GM crops is likely to continue. Therefore, accurate and reliable methods for the detection of GMOs are crucial for regulatory compliance, labeling, and consumer confidence. Polymerase chain reaction (PCR) is a powerful and widely used molecular biology technique that has revolutionized the detection of GMOs due to its sensitivity, specificity, and speed.

In this chapter, we will provide an overview of the PCR-based methods for GMO detection, including the principles, techniques and their limitations, and quality perspectives.

1.2 Basic principles of PCR

Polymerase chain reaction is a widely used technique that allows the amplification of specific DNA sequences in a sample to generate millions of copies, making it highly sensitive for detecting GMOs. The basic principles of PCR involve three main steps: denaturation, annealing, and extension. Denaturation involves heating the DNA sample to separate the double-stranded DNA into single strands. Annealing involves cooling the sample to allow specific primers to bind to complementary sequences in the DNA. Extension involves the addition of a heat-stable DNA polymerase enzyme that synthesizes new DNA strands by extending from the primers. These steps are repeated in cycles, resulting in the exponential amplification of the target DNA sequence. The amplified DNA can then be analyzed using various methods, such as gel electrophoresis, DNA sequencing, or real-time PCR, to detect the presence of GMOs [11, 12].

The detection of GMOs by PCR relies on the use of specific primers that are designed to amplify target DNA sequences that are unique to the genetically modified traits or genes of interest. These primers are usually designed based on the DNA sequences of the genetically modified traits or genes that have been characterized and deposited in public databases. The specificity of the PCR-based GMO detection methods is determined by the specificity of the primers used. The primers should only amplify the target DNA sequence from the GMO of interest and not from other non-GMO organisms or DNA contaminants [13]. The sensitivity of the PCR-based GMO detection methods depends on the region of selection for primers' design, efficiency of the PCR amplification, and the quality and quantity of the DNA sample used. The PCR amplification should be optimized to ensure that even trace amounts of GMO DNA can be detected reliably. It was found that if primers are not from the conserved region, chances are they will miss the target gene (**Figure 1**).

1.3 Protocols for PCR-based GMO detection

The PCR-based GMO detection methods can be categorized into two main types: qualitative and quantitative methods. Qualitative methods determine the presence or absence of GMOs in a sample, while quantitative methods determine the amount or percentage of GMOs in a sample.

1.3.1 Conventional PCR for GMO detection

Conventional PCR, also known as endpoint PCR, is a widely used method for the detection of genetically modified organisms (GMOs). It involves the use of specific primers that target the DNA sequences of the GMO of interest, and these primers are used to amplify the DNA through a series of temperature cycles. The PCR

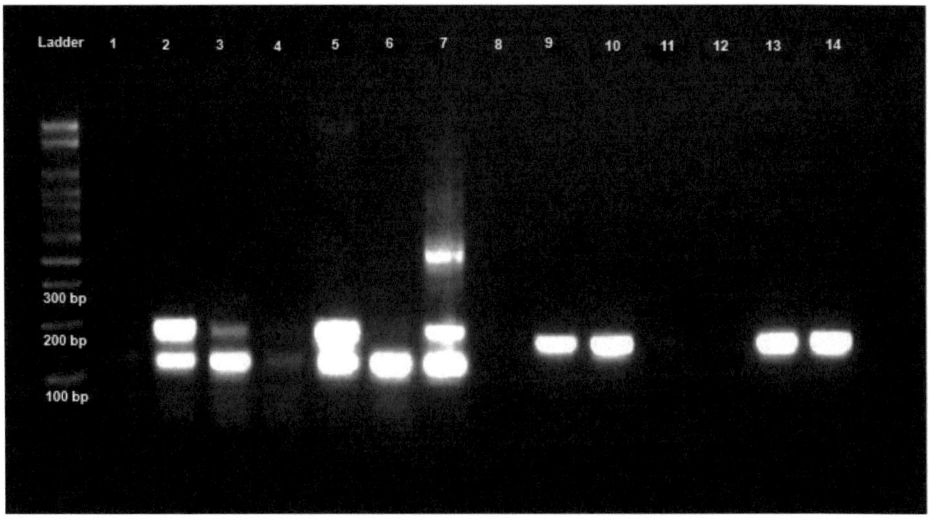

Figure 1.
Duplex PCR for promoter (p35S, 195 bp) and terminator (tNOS, 125 bp) region difference in primer sequence, created difference in results. Key: 1 = blank; 2 = indigenous positive control for tNOS and p35S); 3–11 = Rice samples suspected for genetic modification; 12 = negative control; 13–14 = known samples of rice positive for p35S.

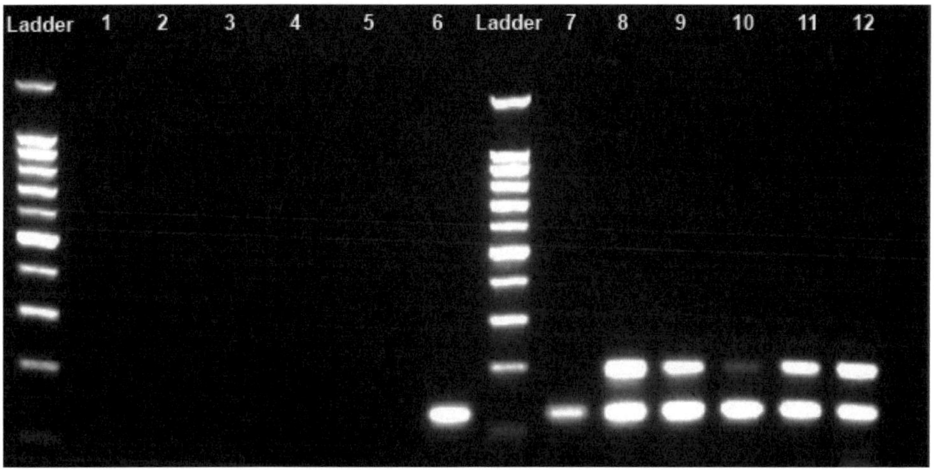

Figure 2.
Difference between 02 set of primers used for same rice samples in a duplex PCR for promoter (p35S, 195 bp) and terminator (tNOS, 125 bp) region. Key: 1–6 = six rice samples suspected for genetic modification tested by one set of primers designed for less conserved sequence; 7–12 = same samples as 1–6 tested by another set of primers designed for conserved regions of promoter and terminator region.

amplification results in the generation of DNA fragments, which can be visualized using gel electrophoresis. The presence of a specific band of the expected size indicates the presence of the GMO in the sample, while the absence of such a band indicates its absence. Duplex (two targets) or multiplex (more than two targets) PCR can also be performed, by designing specific primer pairs that target the DNA sequences of different GMOs of interest. These primers are then used in a single PCR, along with a DNA template extracted from the sample being tested. The PCR amplification results in the generation of DNA fragments that are specific to each GMO, which can be separated and identified using gel electrophoresis or other detection methods.

Duplex/multiplex PCR not only saves time and resources by combining multiple GMO detections into a single reaction, but also provides high sensitivity and specificity, making it a valuable tool in GMO detection and identification in various agricultural, food, and environmental samples (**Figure 2**) [14].

1.3.2 Advantages of conventional PCR

This technique is straightforward to grasp and implement, yielding fast results. It is an extremely responsive method capable of generating millions to billions of replicas of a particular product for sequencing, cloning, and analysis purposes [15].

1.3.3 Limitations of conventional PCR

Conventional PCR can be used for multiplexing as well, so it is not its limitation, however its reliance on gel electrophoresis or other post-PCR methods for DNA fragment analysis is a limitation, as it is an additional step for confirmation which makes conventional PCR time-consuming. Moreover, the high variability of GMO sequences and the need for stringent primer design and validation can also affect the accuracy and reliability of conventional PCR results [15].

1.3.4 Real-time PCR for GMO detection

Real-time PCR, also known as quantitative PCR (qPCR), is a powerful technique used for the detection and quantification of GMOs. Unlike conventional PCR, which provides qualitative results (i.e., presence/absence of a target DNA), real-time PCR allows for the precise quantification of the amount of GMO DNA present in a sample.

Real-time PCR can utilize nonspecific fluorescent dyes like SYBR green fluorescent and labeled probes that specifically bind to the amplification product during the PCR, generating a signal that is proportional to the amount of target DNA.

Real-time PCR allows for real-time monitoring of the amplification process, and the quantification of GMO DNA can be determined based on the threshold cycle (Ct) value, which is the cycle number at which the fluorescent signal crosses a predetermined threshold.

1.3.5 Advantages of real-time PCR

Real-time PCR offers several advantages for GMO detection, including high sensitivity, specificity, and accuracy, as well as the ability to quantify the amount of GMO DNA in a sample. It is widely used in regulatory and commercial GMO testing, as it provides reliable and rapid results for GMO detection and quantification in a wide range of food and feed samples. Multiplex PCR can also be conducted very easily with the help of probes labeled with different fluorescent dyes [15].

S#	Technique	Principal	Advantages	Limitations	Ref
1	DNA staining and hybridization techniques (such as Southern blotting)	Specific detection of target genes through the complementary binding of DNA probes to target DNA sequences	• Simple technique • Specific and very sensitive technique • Does not involve expensive equipment	• This technique may not be sensitive enough to detect low levels of GMO-specific DNA in complex matrices, such as processed foods or feed samples. • Time-consuming technique	[15]
2	Nested PCR technique	It involves the use of two sets of primers in two separate rounds of PCR amplification, resulting in increased sensitivity and specificity compared to conventional PCR.	• Highly specific and very sensitive technique • Can be used for sequences with hypervariable regions • Quantitative evaluation can be done	• Increased risk of contamination • Increased labor and time-consuming • Risk of carryover contamination as two rounds of PCR are required	[15]
3	Capillary gel electrophoresis (CGE)	Multiple PCRs using fluorescently labeled forward primers and discrimination of amplimer of similar magnitude by CGE	• More than eight different targets can be detected in one reaction • Higher resolution as compared to normal PCR multiplex	• Extensive labor in primer designing to avoid the self and cross-complementarity between primers • Sophisticated equipment is required which might not always be available	[16]

S#	Technique	Principal	Advantages	Limitations	Ref
4	Loop-mediated isothermal amplification (LAMP)	LAMP uses multiple primers that recognize distinct regions of the target DNA or RNA sequence. Amplification is done by strand displacement and autocycling strand synthesis, under isothermal conditions.	• Rapid and sensitive detection • Simple and cost-effective • No sophisticated equipment is required • Robustness and tolerance to inhibitors • Visual detection possible	• Nonspecific amplification • Primer design challenge • Limited multiplexing capabilities	[17]
5	Digital PCR (dPCR)	dPCR involves the partitioning of a sample into numerous smaller reactions, each containing a single molecule or a few molecules of the target nucleic acid. The sample is then subjected to PCR amplification, and the endpoint fluorescence is analyzed to determine the presence or absence of the target molecule in each partition.	• Absolute quantification of nucleic acids • Capable of detecting and quantifying rare or low-abundance nucleic acids • Reduces the variability associated with traditional PCR methods, by eliminating the reliance on standard curves and minimizing the effects of PCR inhibitors • The binary nature of dPCR results (positive or negative partitions), reduces the potential for human bias in data analysis	• Expensive and technically challenging compared to traditional PCR methods • dPCR has a limited dynamic range compared to qPCR, which can quantitate a wide range of target concentrations in a single reaction • Optimization of dPCR assays can be challenging due to the need to determine the optimal partitioning strategy, including the number and size of partitions, and the concentration of target molecules in each partition	[16]
6	Microarray technology	Microarray-based gene detection involves the immobilization of thousands of gene-specific probes onto a solid surface. The hybridization event is then detected using a variety of methods, such as fluorescent dyes, chemiluminescent tags, or radioisotopes. The intensity of the signal generated is proportional to the amount of target sequence present in the sample, allowing for quantitative measurement of gene expression levels, or detection of specific genetic mutations.	• High-throughput analysis of gene expression or genetic mutations • Detection of a wide range of genes and their expression levels is possible • Quantitative and reproducible results • Flexibility and customization as per target genes/genomes	• Expensive equipment • Limited sensitivity for low-abundance targets • Complexity and data analysis challenges require specific software • Accuracy and reliability of microarray-based gene detection depend on the quality of the probes • Limited dynamic range, meaning that very high or very low expression levels may not be accurately detected	[16]

S#	Technique	Principal	Advantages	Limitations	Ref
7	Next-generation sequencing (NGS)	NGS involves the parallel sequencing of millions of DNA or RNA fragments using sequencing-by-synthesis methods. NGS generates vast amounts of sequence data, which are then processed and analyzed using bioinformatics tools to assemble and interpret the sequence reads, resulting in the identification of genomic or transcriptomic information.	• High-throughput sequencing • Increased speed and cost-effectiveness • NGS enables the identification of rare mutations, structural variations, and other genetic alterations with high accuracy and sensitivity • Versatility and flexibility	• Data analysis complexity • NGS requires complex and time-consuming sample preparation and library construction steps, including DNA or RNA fragmentation, library amplification, and adapter ligation, which can introduce biases and errors	[16]

Table 1.
Other polymerase chain reaction (PCR)-based genetically modified organism (GMO) detection techniques and their limitations.

1.3.6 Limitations of real-time PCR

Although real-time PCR is a powerful and widely used technique for GMO detection, it also has some limitations that should be considered. Like any PCR-based method, real-time PCR can be susceptible to false-positive or false-negative results. Contamination during sample preparation, amplification, or detection steps, as well as issues related to primer design and specificity, can lead to inaccurate results. Additionally, factors, such as PCR efficiency, DNA extraction efficiency, and sample matrix effects, can further contribute to variability in quantification. Real-time PCR can be more expensive compared to conventional PCR, as it requires specialized equipment, fluorescent probes, and other reagents. The cost of reference materials for quantification can also be high, which may be a consideration for laboratories with limited budgets (**Table 1**) [15].

2. Sample preparation and DNA extraction for PCR-based GMO detection

2.1 Importance of sample preparation and DNA extraction

Contaminants, inhibitors, and degraded DNA in the sample can adversely affect the sensitivity, specificity, and reproducibility of GMO detection. Thorough sample preparation minimizes the risk of false-positive or false-negative results, ensures the integrity of the DNA template, and allows for the efficient and specific amplification of GMO targets. There are many ways to extract DNA from an organic source like:

- Organic extraction (phenol-chloroform method)
- Nonorganic method (salting out and proteinase K treatment)
- Adsorption method (silica-gel membrane)
- Cetyltrimethyl ammonium bromide (CTAB) method, etc.

The GMO plants/crops can have PCR inhibitors, or there may be very trace amount of material which is genetically modified. Among different methods tested, the most promising results with high-through put concentration were found by CTAB method (as described below) [18–20].

2.2 DNA extraction by CTAB method

2.2.1 Sample preparation

Store the samples prior to and during the sample preparation under adequate conditions (e.g., at room temperature, refrigerated, or frozen) depending on the sample.

Prepare all the samples under conditions avoiding cross-contaminations and minimizing degradation of the DNA in the test portion. Seed samples coated with pesticides may be washed to avoid any contamination of GMO dust in the sample, subsequently leading to false-positive results.

Take 60 to 80 mg of sample in a tube and add 15–20 Magnetic Beads to it. Add 01 to 02 ml of CTAB. Set the tubes in for homogenization by a bead ruptor, or manually homogenize the sample by mortar and pestle, any commercial grade grinder can also be used at this step, but it is important to note that this step can be a source of contamination, so utmost care is required in using the mortar and pestle or grinder again with another sample.

Note: Washing with 1 N HCL helps in removing the residual samples from mortar and pestle or grinder.

2.2.2 DNA extraction

Take the freshly homogenized sample (the sample can also be stored at 4°C after homogenization for 1 day). Add 10 μl of proteinase K in the homogenized mixture of plant sample and CTAB. Incubate the tube at 65°C for 30–40 min. After incubation, spin the CTAB/plant extract mixture at 12000 rcf for 4 min and transfer the supernatant into a new tube. Separate the supernatant in a new tube, now add 1.0 μl of ribonuclease (RNase). Incubate the tubes at 37°C for 35 min. Add 200 μl of phenol-chloroform-isoamyl alcohol (PCI) and mix the contents slowly by inverting the tube for 5–7 times. Centrifuge each sample at 12,000 rcf (relative centrifugal force) for 4 min. Wash each sample with 500 μl of 100% ethanol. Wash each sample twice with 650 μl of ethanol (70%). Add 100 μl of Tris EDTA (ethylenediaminetetraacetic acid) (TE) buffer and collect eluant in a new tube by centrifugation for 4 min at 12,000 rcf. Add 1 μl of RNase A and and incubate for 15 min at 37°C. Take optical density (O.D.) of the sample as mentioned in Method ID: BRC/MBD/17025/WI-2.4. After taking the O.D, store the sample at ultra-low temperature, until further use [18–20].

2.3 Estimation of nucleic acid

Quantity of extracted nucleic acid per milliliter (ml) or microliter (µl) can be estimated by a spectrophotometer or qubit (quantum bit) as per standard protocols. The purity of DNA should be between 1.8 to 2.0.

The following website offers a calculator for determining the number of copies of a template DNA (http://www.uri.edu/research/gsc/resources/cndna.html).

The total number of copies of double-stranded DNA may be calculated using the following equation:

Number of copies of DNA = (DNA amount (ng) x 6.022x1023) / (length of DNA x 1x109 ng/ml x 650 Daltons).

Calculating the number of copies of DNA is used to determine how much template is needed per reaction [21].

2.4 PCR

There are certain common factors to consider, regardless of which PCR technique is to be followed. The volumes of a PCR mix vary between 10 and 100 µl. The final concentration of deoxynucleoside triphosphates (dNTPs) should be 200 µM. The primer concentration can vary between 10 and 50 pmol per reaction. The DNA template concentration should be 10 and 500 ng (nanograms) per reaction. The amount of DNA polymerase should not exceed 3 units per reaction.

For calculating the melting temperature ™ values, one of the following calculators is recommended [22]:

http://www6.appliedbiosystems.com/support/techtools/calc/
http://www.cnr.berkeley.edu/~zimmer/oligoTMcalc.html
http://www.idtdna.com/analyzer/Applications/OligoAnalyzer/
http://www.finnzymes.com/tm_determination.html
http://mobyle.pasteur.fr/cgi-bin/portal.py?#forms::melting

The optimal duration and number of temperature cycles in PCR depend on factors, such as the size of the target sequence and the characteristics of the primers, including size and complementarity.

It is recommended to minimize the durations to save time and reduce the risk of nonspecific amplification. Typically, 30 s is sufficient for denaturation and primer hybridization. For elongation, a rule of thumb is 1 min per kilobase of DNA for regular cycles, and 2 mins per kilobase for the final elongation cycle. The number of cycles, usually ranging from 20 to 40, is inversely related to the abundance of the DNA template.

2.5 Gel electrophoresis

Real-time PCR does not need this step. After conventional PCR (single plex, multiplex, or nested), make a gel of agarose, ranging from an agarose concentration from 0.8% to 2.5% inversely related to the size of the band.

Load the mixtures by slowly dropping into the wells. Avoid making bubbles. Attach the electrical leads of power supply to the gel tank (containing gel and buffer), so that DNA can move toward the anode. Apply a voltage of 60–100 volts. Run the gel, until the gel-loading dye migrates the appropriate distance (normally until the bromophenol blue dye front migrates three-fourths of the way down the gel) [23].

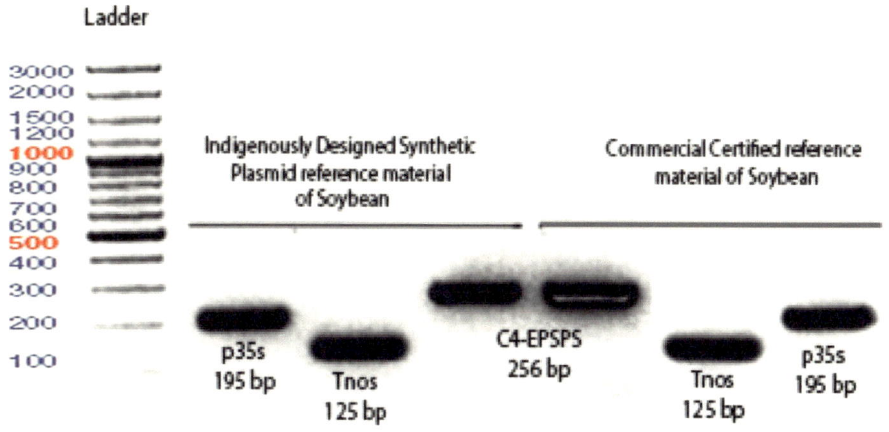

Figure 3.
Indigenously designed reference material of soybean on 2% agarose gel in comparison with commercial certified reference material.

2.6 Troubleshooting of PCR

Detection of unknown GMO DNA in PCR also relies on using reference materials with known GMO content, so that desired band or peak of unknown samples can be compared to the known positive control. However, the availability and quality of reference materials may vary, leading to variability in detection.

This problem was addressed by developing an indigenous reference material at the Centres for Excellence in Science and Applied Technology (CESAT). The indigenous reference material was compared to commercial reference material (**Figure 3**) and it was found that it gave the same results every time.

If the desired amplicon band/peak in positive control along with unknown samples is not obtained under standard PCR conditions, PCR optimization may be necessary to achieve improved results. Adjusting the concentration of $Mg2+$ and/or manipulating annealing temperatures is likely to be effective in improving PCR stringency. However, before making any changes, it is important to rule out human error as the cause of any erroneous results. Confirm that all reagents were properly added to the reaction and were not contaminated.

Additionally, carefully observe the results and ask questions such as whether primer dimers are visible on the gel after electrophoresis, if there are nonspecific products migrating at different sizes, or if there is a lack of any product. Consider whether the target DNA is on a plasmid or in a genomic DNA extract. It is also recommended to analyze the G-C content of the desired amplicon [22–26].

3. Validation and quality control of PCR-based GMO detection methods

Validation and quality control are critical steps in the development and implementation of PCR-based methods for the detection of genetically modified organisms (GMOs). Validation of PCR-based GMO detection methods involves the systematic evaluation of the method's performance characteristics to determine its fitness for the intended purpose. This includes assessing the specificity, sensitivity, accuracy, and robustness of the method [7, 27–29].

3.1 Requirements for validation of GMO detection methods

Validation of GMO (genetically modified organism) detection methods requires meeting certain requirements to ensure the accuracy and reliability of the results. These requirements include establishing the extraction efficiency of the method for each specific matrix being validated, determining the sensitivity (limit of detection-LOD) of the method using appropriate samples, verifying the specificity (selectivity) of the method through theoretical and experimental tests, assessing false-negative and -positive rates, and evaluating the robustness/ruggedness of the method.

Additionally, validation should be conducted using relevant reference materials and following established guidelines or standards, such as those provided by regulatory authorities or international organizations. It is important to thoroughly validate GMO detection methods to ensure their suitability for detecting the presence or absence of GMOs in different samples and to provide reliable results that can be used for regulatory compliance, labeling, and other purposes [7, 27–29].

3.2 Performance parameters

The essential performance characteristics for qualitative PCR methods include extraction efficiency, sensitivity (limit of detection-LOD), specificity (selectivity), false-negative and -positive rates, and robustness/ruggedness.

Extraction efficiency can be determined empirically by testing the extraction method for its efficiency in each matrix being validated. This is necessary to ensure that the extraction is sufficient and reproducible. Known amounts of the target analyte can be spiked into the matrix prior to extraction, and extraction blanks should be included to prevent cross-contamination during the extraction protocol.

Sensitivity can be determined by testing the method at different concentrations of the target sequence. Limits of detection (LOD) should be defined using samples with single ingredients only. The LOD is typically understood as the concentration of the target DNA at which an amplification product is detected with a probability of at least 0.95 (LOD95%). This can be determined by performing a dilution series of the target DNA and conducting PCR replicate measurements at each dilution level. The dilution level with the lowest number of copies for which all replicates are positive is considered an approximate value for LOD95%. The LOD can be expressed as DNA weight/reaction or target copy number/reaction. After determining the LOD of the assay using a dilution series of the target DNA, the originating laboratory should also estimate the LOD of the assay in various food matrices.

Specificity of the method should be tested with DNA/RNA from closely related or potentially co-occurring nontarget species/varieties, as well as DNA/RNA from the reference species/variety material. Verification of specificity can be accomplished through theoretical tests, such as computer-aided ("in-silico") tests, which examine the oligonucleotide sequences (primers, probes) and amplicon sequences for similarities to other sequences by searching suitable databases (e.g. BLASTn). Experimental tests for specificity should also be conducted, including testing with DNA/RNA from nontarget species/varieties (exclusivity) and DNA/RNA from the reference species/variety (inclusivity) material. This testing should include closely related materials and cases where the limits of sensitivity are truly tested [7, 27–29].

3.3 Quality control measures for reliable GMO detection results

Quality control measures are implemented to ensure that the PCR-based GMO detection methods are performing optimally and generating reliable results. This includes monitoring and controlling the quality of reagents, equipment, and experimental conditions, as well as conducting regular checks on the performance of the method.

Quality control measures may include positive and negative controls, amplification controls, and reference materials. Positive controls are samples known to contain the target GMO, while negative controls are samples known to be free of the target GMO. Amplification controls are included to ensure that the PCR amplification step is working properly. Reference materials are commercially available certified reference materials or reference samples with known GMO content that can be used to validate the performance of the method. All this should be used with established standard operating procedures (SOPs) for the PCR-based GMO detection methods and to adhere to them consistently.

Furthermore, interlaboratory validation (ILV) and proficiency testing (PT) programs can be used to assess the performance of PCR-based GMO detection methods across different laboratories. ILV involves the evaluation of the method's performance in multiple laboratories, using the same method and protocol. PT involves the testing of proficiency of laboratories in detecting GMOs through blind testing with unknown samples. ILV and PT programs can help identify any variability or discrepancies in the results obtained by different laboratories, and provide feedback for improving the method and ensuring consistent and reliable results across different laboratories [7, 27–29].

4. Conclusion

The PCR-based techniques are valuable tools for enhancing GMO detection, each technique has its advantages and limitations. It is important to carefully optimize and validate these techniques to ensure reliable and accurate GMO detection results. It may be necessary to use these techniques in combination with other PCR-based methods or analytical approaches to overcome these limitations and obtain robust GMO detection results in different types of samples.

Additionally, regulatory considerations may vary by region or country. Compliance with these regulations, including validation of methods, use of approved reference materials, and adherence to reporting requirements, can add complexity and time to the GMO detection process. Despite these limitations, PCR remains a valuable tool for GMO detection, and careful optimization and validation measures can help mitigate these challenges and ensure reliable results.

These measures include validating the method's performance characteristics, implementing quality control measures, establishing SOPs, and participating in ILV and PT programs. By following these validation and quality control measures, PCR-based GMO detection methods can provide reliable and robust results, which are essential for regulatory compliance, trade, and consumer confidence in GMO detection.

Acknowledgements

The authors acknowledge the Department of Plant Protection, Pakistan, for providing samples for this work.

Conflict of interest

The authors declare no conflict of interest.

A. Appendix – A (Preparation of solutions)

A.1 CTAB buffer preparation

Analytical balance and pH meter are required for CTAB preparation by following reagents: cetyltrimethyl ammonium bromide (CTAB), polyvinylpyrrolidone (PVP40), Tris base, ethylenediaminetetraacetic acid (EDTA), and NaCl.

To prepare CTAB, first prepare solutions A, B, C, and D as follows:

Procedure to make CTAB:

A.1.1 Solution A

Take 2 g of CTAB and 01 g of PVP40. Add in 40 ml of distilled autoclaved water and stir the solution on a hot plate stirrer. Solution A is prepared.

A.1.2 Solution B

Take 2.4 g of Tris base and add in 10 ml of distilled autoclaved water. Adjust the pH to 8.0 and make a final volume up to 20 ml. Solution B is prepared.

A.1.3 Solution C

Take 1.86 g of EDTA and add in 05 ml of distilled autoclaved water. Adjust the pH to 8.0 and make a final volume up to 10 ml. Solution C is prepared.

A.1.4 Solution D

Take 8.475 g of NaCl and add in 30 ml of distilled autoclaved water and dissolve. Solution D is prepared.

For final formulation of CTAB buffer, take 40 ml of solution A, 10 ml of solution B, 4 ml of solution C, and 28 ml of solution D. Gently mix, the total volume of the solution becomes 82 ml. Adjust the pH to 5.0 and make up volume 100 ml with distilled autoclaved water.

A. 2 Tris EDTA (TE) buffer preparation

Analytical balance and pH meter are required for Tris EDTA (TE) buffer preparation by following reagents; Tris (10 mM), EDTA (1 mM), and hydrochloric acid (to adjust the pH).

Procedure to make TE:

Weigh 605 mg of Tris base and 372.24 mg of EDTA in a flask. Dissolve in 400 ml of distilled and deionized water. Adjust the pH to 8.0 with 1 N HCl. Make up the volume until 500 ml. Filter and autoclave the solution for use in downstream processes.

B. Tris acetate EDTA (TAE) buffer preparation

Analytical balance and pH meter are required for Tris EDTA (TE) buffer preparation by following reagents: Tris, acetic acid, EDTA, hydrochloric acid (to adjust the pH).

Procedure to make TAE:

Weigh 121.1 g of Tris base and 9.30 g of EDTA in a flask. Dissolve in 300 mL of distilled and deionized water. Adjust the pH to 8.0 with 28.5 ml of glacial acetic acid. Make up the volume until 500 ml. Filter and autoclave the solution for use in downstream processes.

Author details

Nadia Jamil*, Waqar Hassan Naqvi, Shezaib Siddiqui, Ayaz Ahmed, Ali Dino, Maliha Wajeeh, Zohaib Khan and Saifullah Khan
Centres for Excellence in Science and Applied Technology (CESAT), Islamabad, Pakistan

*Address all correspondence to: diaraj@hotmail.com

References

[1] Eckerstorfer MF, Engelhard M, Heissenberger A, Simon S, Teichmann H. Plants developed by new genetic modification techniques—Comparison of existing regulatory frameworks in the EU and non-EU countries. Frontiers in Bioengineering and Biotechnology. 2019;7:26. DOI: 10.3389/fbioe.2019.00026

[2] Turnbull C, Lillemo M, Hvoslef-Eide TAK. Global regulation of genetically modified crops amid the gene edited crop boom – A review. Frontiers in Plant Science. 2021;**12**:630396. DOI: 10.3389/fpls.2021.630396

[3] Available from: www.fda.gov

[4] National Academies of Sciences, Engineering, and Medicine. Genetically Engineered Crops: Experiences and Prospects. Washington, DC: The National Academies Press; 2016. DOI: 10.17226/23395

[5] Garcia-Alonso M, Novillo C, Kostolaniova P, Parrilla MM, Alcalde E, Podevin N. The EU's GM crop conundrum did the EU policy strategy to convert EFSA GMO guidance into legislation deliver on its promises? EMBO Reports. 2022;**23**:e54529

[6] National Research Council (US) Committee on Identifying and Assessing Unintended Effects of Genetically Engineered Foods on Human Health. Safety of Genetically Engineered Foods: Approaches to Assessing Unintended Health Effects. Washington (DC): National Academies Press (US); 2004. PMID: 25009871

[7] Zel J, Milavec M, Morisset D, Plan D, Van Den Eede G, Gruden K. How to Reliably Test for GMOs. 2012. DOI: 10.1007/978-1-4614-1390-5

[8] Rozas P, Kessi-Pérez EI, Martínez C. Genetically modified organisms: Adapting regulatory frameworks for evolving genome editing technologies. Biological Research. 2022;**55**:31

[9] Vezzani S. The international regulatory framework for the use of GMOs and products thereof as food aid. European Journal of Risk Regulation. 2018;**9**:120-136

[10] Jones MGK, Fosu-Nyarko J, Iqbal S, Adeel M, Romero-Aldemita R, Arujanan M, et al. Enabling trade in gene-edited produce in Asia and Australasia: The developing regulatory landscape and future perspectives. Plants. 2022;**11**:2538. DOI: 10.3390/plants11192538

[11] Lorenz TC. Polymerase chain reaction: Basic protocol plus troubleshooting and optimization strategies. Journal of Visualized Experiments. 2012;**63**:e3998. DOI: 10.3791/3998

[12] Delidow BC, Lynch JP, Peluso JJ, White BA. Polymerase chain reaction. In: White BA, editor. PCR Protocols. Methods in Molecular Biology. Vol. 15. Totowa, NJ: Humana Press; 1993. DOI: 10.1385/0-89603-244-2:1

[13] Ye J, Coulouris G, Zaretskaya I, Cutcutache I, Rozen S, Madden TL. Primer-BLAST: A tool to design target-specific primers for polymerase chain reaction. BMC Bioinformatics. 2012;**13**:134

[14] Safaei P, Aghaee EM, Khaniki GJ, Afshari SAK, Rezaie S. A simple and accurate PCR method for detection of genetically modified rice. Journal of Environmental Health Science

and Engineering. 2019;**17**(2):847-851. DOI: 10.1007/s40201-019-00401-x

[15] Hajia M. Limitations of different PCR protocols used in diagnostic laboratories: A short review. Modern Medical Laboratory Journal. 2018;**1**:1-6

[16] Salisu IB, Shahid AA, Yaqoob A, Ali Q, Bajwa KS, Rao AQ, et al. Molecular approaches for high throughput detection and quantification of genetically modified crops: A review. Frontiers in Plant Science. 2017;**8**:1670. DOI: 10.3389/fpls.2017.01670

[17] Takabatake R, Kagiya Y, Minegishi Y, Yeasmin S, Futo S, Noguchi A, et al. Development and evaluation of rapid screening detection methods for genetically modified crops using loop-mediated isothermal amplification. Food Chemistry. 2017;**252**:390-396

[18] Berben G, Charels D, Demsar T, Hochegger R, Nardini E, Onori R, et al. Report EUR 27021 EN - Guidelines for sample preparation procedures in GMO analysis. 2014. DOI: 10.2788/337005

[19] ISO-2157:2005(E). Sample preparation procedures in GMO analysis

[20] Nguyen TCT, Son R, Raha AR, Lai OM, Michael CWVL. Comparison of DNA extraction efficiencies using various methods for the detection of genetically modified organisms (GMOs). International Food Research Journal. 2009;**16**:21-30

[21] Foodstuffs. Methods of Analysis for the Detection of Genetically Modified Organisms and Derived Products, Qualitative Nucleic Acid-Based Methods. 1st ed. ISO-21569:2005-6-15

[22] Mammedov TG, Pienaar E, Whitney SE, JR TM, Carvill G, Goliath R, et al. A fundamental study of the PCR amplification of GC-rich DNA templates. Computational Biology and Chemistry. 2008;**32**(6):452-457

[23] Lee PY, Costumbrado J, Hsu CY, Kim YH. Agarose gel electrophoresis for the separation of DNA fragments. Journal of Visualized Experiments. 2012;**62**:3923. DOI: 10.3791/3923

[24] Strien SJ, Mall GJ. Enhancement of PCR amplification of moderate GC-containing and highly GC-rich DNA sequences. Molecular Biotechnology. 2013;**54**:1048-1054

[25] Hubé F, Reverdiau P, Iochmann S, Gruel Y. Improved PCR method for amplification of GC-rich DNA sequences. Molecular Biotechnology. 2005;**31**(1):81-84

[26] Barbau-Piednoir E, Lievens A, Mbongolo-Mbella G, et al. SYBR® green qPCR screening methods for the presence of "35S promoter" and "NOS terminator" elements in food and feed products. European Food Research and Technology. 2010;**230**:383-393. DOI: 10.1007/s00217-009-1170-5

[27] Stromgren SA. Guidelines for the Validation of Analytical Methods for Nucleic Acid Sequence-Based Analysis of Food, Feed, Cosmetics and Veterinary Products. 1.1 ed. USA: FDA; 2020

[28] Stromgren SA. FDA Guidelines for the Validation of Analytical Methods for the Detection of Microbial Pathogens in Foods and Feeds. 3rd ed. US: Food and Drug Administration Foods Program; 2019

[29] Bruetschy C. The EU regulatory framework on genetically modified organisms (GMOs). Transgenic Research. 2019;**28**:169-174

Section 3

Cellular Applications

Chapter 3

The Employment of Polymerase Chain Reaction in Building Phage Displayed Antibody Libraries

Abebe Zemene

Abstract

"The power of evolution is revealed through the diversity of life," says the announcement for the 2018 Nobel Prize in chemistry, George Smith described phage display technology in 1985, and it has since been expanded to include the display of peptides, various recombinant antibody formats, enzymes, and fragmented proteomes. These phage libraries can display a wide range of ligands on coat proteins projecting from the surface of the bacteriophage particle, with the ligand-coat protein fusion encoded using either a phagemid or phage vector system. Most phages have a high affinity for host cell surface receptors, such as receptor binding proteins (RBPs) or lipopolysaccharides (LPS). Recently, a number of real-time PCR assays for the rapid, sensitive, and specific detection of each F-RNA phage genogroup have been developed and RNA is reverse-transcribed into cDNA, used for PCR of the encoded antibodies' VH and VL chains. The human antibody repertoire evolved to recognize any target with sufficient specificity and affinity that used RT-PCR to amplify the VH and VK and Vλ chains. Antibody phage display is a versatile *in vitro* selection technology for identifying high-affinity antibodies specific to a wide range of antigens. Low solubility affects the bioavailability, immunogenicity, and effectiveness of mAbs.

Keywords: phage characteristics, phage display, antibody libraries design, bio-panning, the application of PCR

1. Introduction

"The power of evolution is undeniable in the diversity of life," said the announcement of the 2018 Nobel Prize in Chemistry, awarded to Frances H. Arnold for directed enzyme evolution and George P. Smith and Gregory P. Winter for presenting peptides and antibody. Now that scientists have taken control of evolution, they can use it for the benefit of mankind. Directionally evolving enzymes are used to make everything from biofuels to pharmaceuticals, and antibodies displayed on phages are used to treat autoimmune diseases and cancer. However, new algorithms and advances in computational power have enabled the development of new and improved synthetic protein structures and functions, designed entirely *in silico* using *de novo* and rational molecular design, a combined synergy of computational design and evolutionary approaches. Compared to those made only on a design basis. In this way, protein engineering complements and follows genetic engineering, a multidisciplinary field that organically integrates chemistry, physical and molecular biology, computer science, and proteins [1, 2].

In terms of safety, therapeutic antibodies should exhibit negligible or no immunogenicity in recipients, especially in cancers that require repeated administration in long-term therapeutic protocols; this implies that fully human antibodies are the safest antibody isotype/format, in contrast to animal-derived equivalents such as mouse monoclonal or even humanized animal antibodies. The production of fully human antibody isotypes targeting proteins of human origin, such as B. The human oncoproteins overexpressed by cancer cells, however, is difficult and requires a specific strategy (B cells in peripheral lymphoid tissues have passed selection for bone marrow and are not expected to have B cell receptors for self-antigens [3].

George Smith described phage display technology in 1985, and it has since been expanded to include the display of peptides, various recombinant antibody formats, enzymes, and fragmented proteomes. These phage libraries can display a wide range of ligands on coat proteins projecting from the surface of the bacteriophage particle, with the ligand-coat protein fusion encoded using either a phagemid or phage vector system. In general, small ligands like peptides are usually displayed on either coat protein pIII or pVIII, while larger protein fragments like those derived from antibodies are usually displayed on pIII. Phage display libraries can make it easier to find ligands with desirable binding properties, such as those that bind to toxic or non-immunogenic antigens. While there are commercial sources for both peptide and antibody libraries, the exploitation of any isolated ligand is frequently hampered by library ownership. As a result, it may be necessary to create bespoke synthetic one-pot ligand libraries, containing a wide range of potential binders capable of providing antibodies or peptides to any target antigen with no restrictions on use [4, 5]. The restriction enzyme cuts the vector. The circularized vector is then used for bacterial transformation, with a short single-stranded gap. Kong *et al.* recently described a method for producing high-diversity peptide libraries *via* a whole-plasmid PCR self-ligation process that greatly simplifies and increases the efficiency of library construction [6, 7].

In relation to phage display antibody formats, Fabs are mainly composed of the VH-CH1 and VL-CL chains, one of which is fused to the pIII protein. ScFvs are made up of only the VH and VL chains connected by a peptide linker, with one chain fused to the pIII protein, whereas VHHs are made up of a single heavy chain domain equivalent to a conventional IgG VH domain. VHHs are derived from natural heavy chain antibodies found in camelids that lack a light chain. They are also known as single-domain antibodies or nanobodies. Synthetic antibody libraries are typically built from a single or very few "scaffold" diversifying sequences created into one or more of the complementarity determining regions (CDRs). Conventional cloning of Fab and scFv libraries employs multiple PCRs to introduce diversity into CDR regions, followed by using an overlapping extension PCR to magnify the scFv or to distinguish the VL and VH chains for Fabs. The product lines are therefore cloned using various restriction enzymes complementary with the phage or phagemid vector. For instance, a variation on this strategy is the use of type IIS restriction enzymes to facilitate the so-called "seamless" cloning of CDR regions. Because this type of restriction enzyme cuts outside of its recognition sequence, an overhang compatible with any sequence in the scaffold can be produced. Subsequent cloning results in a DNA sequence devoid of any restriction enzyme site, designed to allow the addition of extra amino acids within expressed sequences to be avoided, and the process is irreversible. Alternative methods use Kunkel mutagenesis to introduce diversity into multiple CDRs at the same time by annealing oligonucleotides (containing diversity) to CDR regions in scFv-phagemid template ssDNA and then synthesizing the second strand of the DNA to produce hetero duplex DNA that is used directly for transformation or is further amplified by rolling circle amplification before transformation [8–10].

Fab libraries are available commercially, such as MorphoSys' Ylanthia library. This library is extremely diverse, with 36 fixed heavy/light chain pairs chosen from 400 combinations for highly desired biophysical properties, such as high expression levels and thermal stability. Limited diversity was introduced into CDRH1 and CDRH2, and high diversity was then introduced into CDRH3 and CDRL3, yielding a library of ~1×10^{11} clones. The authors created oligonucleotides that tightly control the codons inserted in the diverse regions using trimer-phosphoramidite-oligonucleotides or a proprietary gene synthesis technology called Slonomics®. The diversified synthetic gene regions were then cloned into the scaffold with type IIS restriction enzyme cloning. The Tomlinson I and J library, which was created by mutating CDR2 and CDR3 regions of both VL and VH in a single scFv scaffold, and the HuCal library, which introduced diversity into CDR3s, are two examples of scFv libraries. The HuCal library was created by combining 7 VL and 7 VH scaffolds to create 49 combinations into which CDR3 diversity was introduced using trimer-phosphoramidite-oligonucleotides, yielding a library with ~2×10^9 diversity. CDR diversity was cloned using engineered restriction sites within scaffolds flanking the CDR3 domains. An alternative method for creating scFv libraries was to use modular cloning of different CDR2 and CDR3 regions. Heavy and light chain CDR2 and CDR3 sequences found within a specific scFv framework were mined from antibody sequence databases. Based on 2000 sequences, rearranging the four CDR regions resulted in a library with ~1×10^{10} diversity. To introduce the distinct CDR regions into the scaffold clone, Kunkel mutagenesis was used, followed by Rolling Circle Amplification to further amplify products before ligation and transformation [10, 11].

Therapeutic and reagent antibodies can be found in phage display antibody libraries. Due to the small size and solubility of phage particles, as well as the combinatorial nature of most antibody libraries, up to 10^{11}–10^{12} clearly different antibodies could be displayed and carefully selected. Further to that, the phage display method has the potential to tailor *in vitro* screening parameters for clearly delineated antibody characteristics, associated with its high-throughput nature, which tends to make it a very powerful platform. Due to the difficulties in bacterial expression, folding, and display of full-length IgG molecules, phage display antibody libraries and other *in vitro* antibody display methods such as ribosome and mRNA display typically use antibody fragments such as antigen-binding fragments (Fabs) or single-chain variable fragments (scFvs), because of the simplicity of the display vector and the higher expression levels of scFv in *Escherichia coli* (*E. coli*) compared to Fab, scFv phage libraries are especially prominent [12, 13].

Following *in vitro* selection, soluble scFv from single *E. coli* colonies can be expressed in sufficient quantities for high-throughput screening (HTS), reducing the number of scFv antibodies for further characterization. This screening, however, is surrogate in nature and has limitations such as the inability to assay for properties that are dependent on bivalent binding or Fc-mediated function of an antibody, and the tendency of scFv antibodies to aggregate, which results in false positive or negative results because full-length IgG is the most frequently used antibody drug furthermore, endotoxin-sensitive, cell-based functional assays do not work with scFv expressed in bacteria. This is a significant disadvantage because functional assays are typically the most useful in determining the most relevant antibodies. Besides that, these assays frequently actually require the purification of scFv samples, which reduces screen throughput [14, 15].

Immunological memory for specific pathogens and diseases is represented by the repertoire of antigens and epitopes specifically recognized by antibodies. Thus, antigen identification and epitope mapping are both critical steps in developing vaccines, diagnostic tests, and therapeutic options for infectious and parasitic diseases. Chagas disease, leishmaniasis, and malaria, as well as emerging multidrug resistance yeast species (e.g., *Candida auris*), are examples of diseases caused by eukaryotic infectious agents for which there are no vaccines or optimal therapeutics, and which continue to pose a challenge to modern medicine and health systems worldwide. Along with them, the twenty-first century has seen a series of viral pandemics, including the *Zika* and *chikungunya viruses*, as well as respiratory viruses like *influenza H1N1* and *coronavirus*-induced severe respiratory syndromes like severe acute respiratory distress syndrome [16]. Large-scale antibody repertoire mapping, particularly against pathogens with large genomes, can be particularly difficult. Such endeavors have long relied on a diverse set of techniques, ranging from enzyme-linked immune-sorbent assays (ELISA) to cDNA libraries and microarray platforms. All of these technologies are time-consuming and require expensive and specialized reagents: custom arrays actually require the cloning and expression of thousands of proteins or the design and custom-ordering of proprietary peptide arrays. First, genes that have not yet been identified or characterized are not always included in such assays, which is a problem that is frequent in protozoan parasites because the majority of their genes are still classified as hypothetical or predicted. Furthermore, many full-length proteins are not always well expressed, which means they may not fold properly, potentially epitopes that require peptides with lengths that surpass the array's parameters. Finally, once an antigen has been identified, epitope mapping is required, which frequently involves the use of expensive peptide tiling and/or arrays. We show here that phage display combined with computational approaches has the potential to address many of these methodological limitations, allowing for the development of a less cumbersome, more cost-effective, and unbiased platform for simultaneous antigen identification and epitope mapping [17–19].

2. General characteristics of phages

Even though bacteriophages (phages) are so common in crops, live animals, and human intestinal environments, humans come into direct and indirect contact with them. Many studies have found bacteriophages in fermented foods such as yogurt and cheese. Specific bacteriophages added to foods aid in the reduction of pathogenic bacteria. Bacteriophages are bacterial viruses that lyse a susceptible bacterial culture completely. Because most virulent phage replication results in bacterial death, interactions between phages and bacteria can be considered parasitism. Certain interactions are mutualistic, and some temperate phages encode benefits for the host bacteria's phenotypic properties. According to Batinovic *et al.*, bacteriophages have been present in the environment for billions of years, resulting in a balance of commensal and pathogenic bacteria. Phages and bacteria are the most ancient and widespread microorganisms on the planet, having evolved roughly 3 billion years ago [20, 21]. Although bacteriophages can exist independently outside of the host, all phages require a bacterial cell to multiply. Most phages have a high affinity for host cell surfaces receptors such as receptor binding proteins (RBPs) or LPS [22]. Based on their production and generation, phages are classified into three types: lytic phages like T4, temperate phages like Lambda (λ), and lysogenic phages like M13. M13 is a filamentous phage that transforms the host cell into a factory without causing lytic disruption. This phage can escape by leaking through the bacterial membrane without disrupting the cell wall or membrane. In contrast, lytic phage infections cause bacterial cells to be lysed and disrupted by viral replication within the bacterial cell, and the infection spreads quickly to new hosts. Temperate phages, the third type, can develop *via* either a lytic or a lysogenic pathway. The phage genes in temperate phages become integrated into the bacterial genome. The phages replicate within the natural bacterial cell cycle and are transmitted to the next generation of bacteria; however, under stressful conditions such as UV irradiation, the phages become activated and the lytic phase begins [23].

A polypeptide or an antibody fragment, such as a scFv, can be expressed on the phage surface. While other *in vitro* expression methods, such as yeast surface display, ribosome display, and puromycin display are readily accessible, the phage display system is the most commonly used. The phage used to display antibodies is the heart of phage display technology. A wide range of phages, including filamentous T4 and M13, can be used in the phage display system. The M13 phage is most commonly used. The filamentous phage is characterized by a filamentous structure and a circular single-stranded DNA (ssDNA) genome. The most common filamentous phages infecting *E. coli* are *M13, f1* and *fd* phages, all of which are Ff phages. The ssDNA genome of the Ff bacteriophage is approximately 6400 bp long and encodes 11 different genes. These genes are classified as structural proteins (genes III, VI, VII, VIII, and IX) and functional proteins (genes I, II, IV, V, X, and XI), which are required for phage replication and assembly are. The classification of the proteins into those that initiate phage replication (+=upstream) and (-=downstream) from the source of replication is obvious and accurate. The ssDNA genome of the Ff virion is completely enclosed in a cylinder of 2700 molecules of the major protein pVIII. The asymmetric end of Ff phage contains 5 secondary coat proteins, pIII and pVI at one end and pVII and pIX at the other. The ability to replicate but not kill the host bacterium characterizes the Ff bacteriophage. After infection of the host bacterium, the Ff phage replicates as an extrachromosomal element instead of integrating into the host chromosome, allowing the host bacterium to continuously shed virus particles. Consequently, M13 and its related Ff phage are preferred over lytic phage (such as T4 and T7). Phage M13 is the most commonly used vector for phage display due to its high replication capacity and its ability to take up large amounts of foreign DNA. Infection of the host bacterium is achieved by binding the pIII phage to the F′ fimbriae of *E. coli*. Once in the bacterium, ssDNA is converted to double-stranded DNA by host enzymes, resulting in ssDNA and phage protein. Phage assembly occurs in or near the inner membrane of the host bacterium and in the periplasm. Each M13 particle contains 3–5 copies of the pIII protein, with 2700 copies of the most abundant coat protein, pVIII [24] (**Figure 1**).

2.1 Filamentous phages

Filamentous Inoviridae phages are typically about 900 nm long and only 7 nm wide. In this section, we will look at strains from the Ff family, such as M13 and fd. The major coat protein pVIII is present in 2700 copies and forms the side wall that

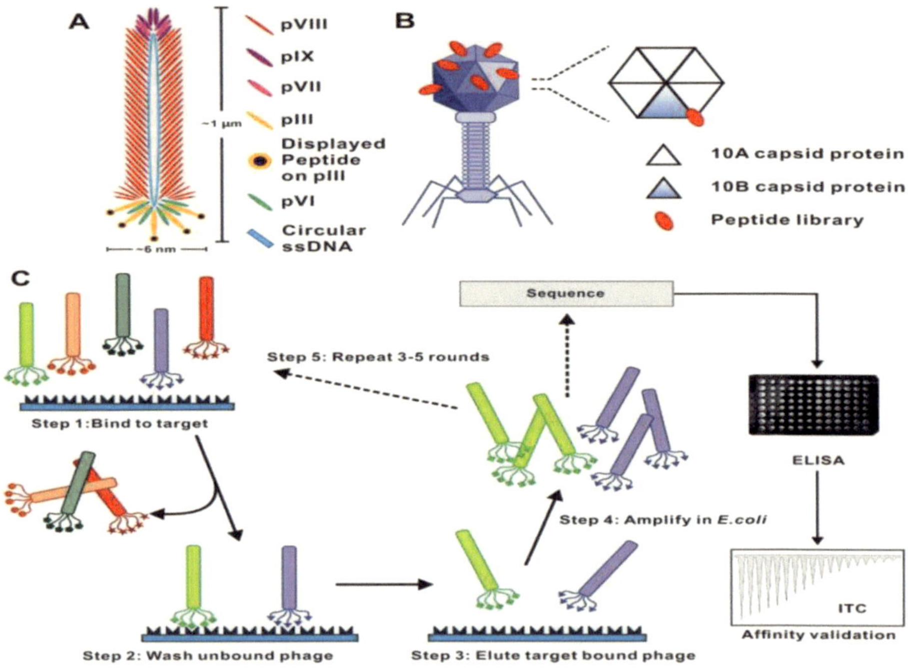

Figure 1.
A schematic representation of phage vector structures and the general bio-panning procedures. Source: Zhang et al. [25]; with permission.

protects the circular 6.4 kb ssDNA genome. Furthermore, the phage has two minor coat proteins, pIII and pVI, on one end and two other minor coat proteins, pVII and pIX, on the other. This phage family only infects *Escherichia coli* (*E. coli*) with a F pilus. pIII's N-terminus first binds to the F pilus, allowing viral DNA to enter the cytoplasm. As the cellular machinery generates progeny genomes, they exit the cell envelope and acquire coat proteins. Importantly, the host bacterium can continuously produce and secrete filamentous phage without being killed. This allows for the rapid and sustained production of phages, which could be useful for developing vaccines against emerging threats. Another advantage of the filamentous phage display system is that antigens can be displayed by all five coat proteins. It is widely used in the field and is regarded as one of the most efficient display systems. Due to its high copy number, pVIII is most commonly used, but it can only display short peptides because it must be extruded through the cell envelope [26, 27].

2.2 T4 and T7 phages

T4 is a *Myoviridae* lytic phage strain with a tail and tail fibers, as well as an elongated icosahedral head measuring approximately 120 nm by 86 nm. Major capsid protein gp23, vertex protein gp24, portal protein gp20, small outer capsid protein (Soc), and highly antigenic capsid protein are phage capsid proteins (Hoc). T4 phage genome is 171 kb dsDNA that exits through the portal vertex formed by gp20 to infect bacteria. The phage enters a vegetative state after infection and must lyse the cell to release viral progenies. Soc forms a cage around the phage's head to protect it from extreme heat, pH, and enzymatic degradation, which is beneficial not only for vaccine administration within the human body but also for transport to developing economies. Both the Soc and Hoc proteins have been identified as antigen display proteins, with the N- and C-termini of the 870 copies of Soc capable of displaying peptides while still not interfering with immune cell recognition. Because the binding site is at the C-terminus of the 155 copies of Hoc, the N-terminus is more desirable for peptide display. The Hoc protein has also been shown to facilitate binding to mammalian cells, potentially allowing for improved vaccine delivery properties [28]. T7 is a lytic phage of the Podoviridae family with a 56 nm icosahedral head composed of two major capsid proteins, gp10A and gp10B, and a 40 kb dsDNA genome. The phage particle also contains the connector protein gp8, the tail proteins gp11 and gp12, and the tail fiber protein gp12. The 29 nm tail promotes bacterial binding, and the proteins gp15 and gp16 encourage DNA insertion. To release new phage particles, the host cell, like T4, is lysed. Because gp10B is not required for capsid assembly, it has been studied for C-terminal antigen display with small peptides in high copy numbers or larger proteins in low to mid copy numbers. Interestingly, it has been demonstrated that the density of antigens displayed on particulate materials influences the type of immune response that develops, implying a delicate balance between using phages to display antigens in low or high copy numbers [29].

2.3 Lambda (λ) phages

λ phage, a Siphoviridae strain, has a 60 nm icosahedral head with a 4 nm shell thickness and a flexible tail about 150 nm long. The major capsid proteins gpD and gpE, the portal protein gpB, the scaffolding protein gpNu3, the viral protease gpC, and the major tail protein pV make up the phage. Interestingly, λ is a temperate phage that can enter a vegetative state of growth or a lysogeny state in which the genome is stored in the cell after cell infection. For DNA vaccination strategies, the large 48.5 kb linear dsDNA genome is commonly used. Furthermore, because the phage contains up to 420 copies of the gpD protein, both the N- and C-termini are available for fusion. Furthermore, the C-terminus of the pV tail protein can be utilized for display, designed to allow for the fusion of multiple antigens per phage particle [30].

2.4 Qβ phages

Qβ phage, such as MS2 phage, has a short ssRNA genome (4.2 kb) with only a few proteins: the major coat protein CP, the maturation protein A2, and the replicase. The 28 nm icosahedral phage also binds to the F pilus in order to infect host cells, where A2 inhibits cell wall synthesis, resulting in lysis and the release of new Qβ phage particles. Due to a leaky stop codon, the A1 protein, which contains a 196-amino acid extension of the C-terminus of CP, is occasionally produced. The read-through domain of A1 is required for host cell infection and is the most frequently used site for antigen display. The sections that follow provide several examples of phage display in infection and cancer. To begin, we will discuss phage-based vaccines that contain antigens associated with parasites, fungi, viruses, and bioterrorism agents [31].

2.5 MS2 phages

MS2 phage originally belonged to the Leviviridae family, which includes the smallest and most fundamental phages. The size of antigen-delivering particles can affect their trafficking to infection sites and immune tissues, as well as the response

Phage	**Filamentous**	**T4**	**T7**	**Λ**	**MS2**	**Qβ**
Structure	Filamentous	Icosahedron + tail			Icosahedron	
Family	*Inoviridae*	*Myoviridae (lytic)*	*Podoviridae (Lytic)*	*Siphovi-ridae*	Levivir ida	*Levivir-idae*
Major dimension	900 × 7 nm	120 × 86 nm	56 × 29 nm	60 × 150 nm	26 nm	28 nm
Vaccine Applications	• Parasites • Fungi • Influenza • Melanoma • Breast cancer	• Anthrax • Plague • Lung cancer	• Influenza • FMD • Epstein–Barr Melanoma Breast cancer	• Hepatitis B • Breast & Liver cancer	• FMD • HPV Zika	Breast cancer Melan-oma
Genome	6.4 kbp ssDNA Major coat protein pVIII	171 kbp dsDNA	40 kbp dsDNA	48.5 kbp dsDNA	3.57 kbpssRNA smallest & simples	4.2 kbp ssRNA

Table 1.
The structure and applications of the phage families employed in vaccine designs.

to those results. The icosahedral size of this small phage family of 26 nm may allow for rapid draining to lymph nodes where displayed antigens can be displayed to immune cells. The 3.57 kb ssRNA genome encodes the major coat protein CP, the maturation protein A, the replicase, and the lysis protein L. MS2 binds to the *E. coli* F pilus, comparable to filamentous phage, to insert its genome. However, unlike filamentous phage, the release of new viral particles necessarily requires host cell lysis. While both the C and N-termini are considered necessary for the arrangement of the 180 copies of CP, an exposed β-hairpin has been employed for antigen display (**Table 1**) [32].

3. The applications of PCR in antibody amplification

Antibody phage display (APD) is a technique that combines bacteriophage genetic engineering with multiple rounds of antigen-guided selection and phage propagation. This technique enables the *in vitro* selection of mAbs of virtually any specificity, greatly facilitating the recombinant production of reagents for use in research and clinical diagnostics, as well as pharmaceuticals for human therapeutic use. The creation of an antibody library is the first step in the APD process. Next, variable heavy (VH) and variable light (VL) PCR products are ligated into a phage display vector, and then mAb clones are analyzed. To isolate specific mAbs from a cloned immunoglobulin repertoire, a large antibody library and efficient selection are required. The preparation of high-quality RNA from the chosen cell source is critical to success. This RNA is reverse-transcribed into cDNA, which is used for PCR of the encoded antibodies' VH and VL chains. Defined primer sets specific for the various VH and VL chain-region gene families then allow amplification of all transcribed rearranged variable regions within a given immunoglobulin repertoire for library construction, thus reflecting all antibody specificities in a specific individual. This makes recombinant cDNA clones for expressed Igs immortal. Among the variations on phage display libraries are (i) libraries designed for Ig isotypes (e.g., IgG, IgA, and IgE) and (ii) libraries of mAbs expressed as Fab fragments or single-chain variable fragments (scFv), the latter of which consists of the VH and VL joined by a linker. Because of the physical connection established between the antibody fragment on the outside and the genetic information encoding the displayed protein within the phage, APD also allows for comprehensive studies of the genetics and function of antigen-specific mAbs. The APD procedure starts with the preparation of an antibody library, followed by the ligation of the variable heavy (VH) and variable light (VL) PCR products into a phage display vector, and finally the analysis of mAb clones.

APD also enables comprehensive studies of the genetics and function of antigen-specific mAbs due to the physical connection established between the antibody fragment on the outside and the genetic information encoding the displayed protein within the phage. The APD procedure starts with the preparation of an antibody library following the ligation of variable heavy (VH) and variable light (VL) PCR products into a phage display vector, mAb clones were analyzed. Recently, a number of real-time PCR assays for the rapid, sensitive, and specific detection of each F-RNA phage genogroup have been developed. However, there is some concern about the applicability of these newly developed real-time PCR assays to genotyping of F-RNA phages in other regions or countries around the world because these assays were designed using a very limited number of F-RNA phage sequences from the DNA database [33].

4. Types of antibody repertoires (libraries)

While library size can be defined by a single number (N) and effective size by the percentage of antibody display, library diversity is more difficult to define. The size of an antibody repertoire can sometimes be used to characterize its diversity. It does, however, assume that the antibody repertoire is composed of a huge array of antigen-binding site shapes and amino acid side chains. Furthermore, not all antibody repertoires are the same. For instance, if a library of antibody variants were created by diversifying all 11 positions of the LCDR1 with 20 amino acids, the resulting repertoire would be 20^{11} or 2×10^{14} antibody variants. Based on its size, this appears to be a diverse repertoire with such a significant number of functioning antibodies. Even though, the majority of the amino acids in LCDR1 are located on the outskirts of the antigen-binding site and play only a minor role in determining antibody binding properties, particularly for smaller and medium-sized targets. As a result, a library of this category would almost certainly be of limited use. In contrast, the same library size with 11 positions mutated to the 20 amino acids in the HCDR3, which is located at the center of the antigen-binding site and plays a crucial role in recognizing diverse targets, would likely yield antibodies with reasonable affinity. As a result, diversity is defined as the number of functional molecules capable of recognizing as many diverse targets as possible, as well as the number of in-frame and/or nontoxic antibody variants. The first functional antibody repertoires were obtained from immunized animals or humans and used as a substrate to build phage display antibody libraries. Immune libraries have had limited application as general-purpose platforms for human therapeutic antibody discovery. Because the libraries are biased toward the recognition of the antigen used as an immunogen, those obtained from immunized animals are single-use libraries.

More importantly, because antibodies derived from immune libraries are nonhuman proteins, they must be further engineered (humanized), raising the cost of antibody drug development. Immune human antibody libraries, for ethical reasons, have mostly been used to isolate antibodies against infectious diseases. The first functional antibody repertoires were obtained from immunized animals or humans and used as a substrate to build phage display antibody libraries. Immune libraries have had limited application as general-purpose platforms for human therapeutic antibody discovery. Because the libraries are biased toward the recognition of the antigen used as an immunogen, those obtained from immunized animals are single-use libraries. More importantly, because antibodies derived from immune libraries are nonhuman proteins, they must be further engineered (humanized), raising the cost of antibody drug development. Immune human antibody libraries, for ethical reasons, have mostly been used to isolate antibodies against infectious diseases. Nevertheless, recent studies of the immune repertoire of Camelid V regions have revealed identical CDR conformations to those found in the human V germline gene repertoire. Because of this similarity, Argen X has created a discovery platform based on camelid immunization. Argen X has discovered several promising therapeutic antibodies that are difficult to obtain through other means, with some of them currently in advanced clinical trials (https://www.argenx.com/en-GB/content/argenx-in-short/2/) [34–36].

In the early 1990s, the first human universal or general-use library was published. This library was established by using repertoire of human genes encoding the circulating antibodies. Because they are not biased toward any particular target, libraries derived from this source of antibody genes are referred to as naive. Although successful, naive libraries contained antibody genes that were toxic to *E. coli,* which,

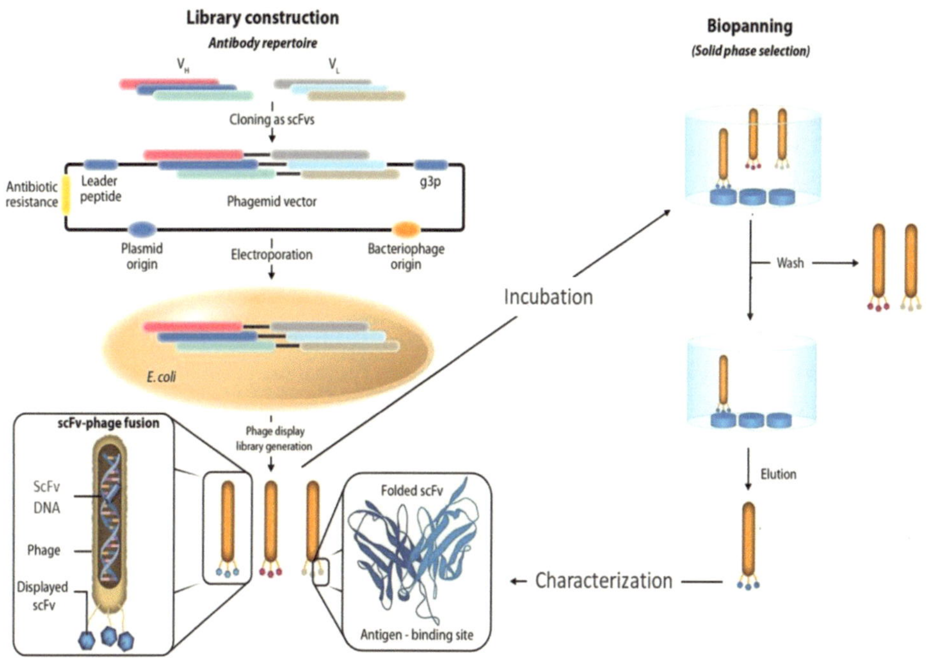

Figure 2.
Phage display methodology that consists of building a library (left panel) of peptide or protein variants or in the case of antibodies, an antibody gene repertoire and affinity selecting (right panel) specific antibody-phage fusions via affinity with the target of interest. Source: Almagro et al. [38]; with permission.

as previously discussed, reduced the effective size of the library. To address this limitation, synthetic antibody libraries were developed. The libraries were carefully designed utilizing assumptions on the number of scaffolds, positions to diversify, types of amino acids to include in the design, and the proportion of each amino acid per position to diversify in this alternative method. These assumptions were not always correct, particularly when it comes to the HCDR3. To avoid making assumptions about the structure of the CDRs, semisynthetic libraries were constructed by combining synthetic scaffolds and natural CDRs [37] (**Figure 2**).

4.1 Naïve libraries

In naive libraries, the diversity of the antibody repertoire is not assumed. According to rationality, the human antibody repertoire evolved to recognize any target with sufficient specificity and affinity. As a result, the goal of working to develop naive libraries is to replicate the diversity of the human antibody repertoire while avoiding biases and redundancy caused by the immunological history of a few individuals and/or rare polymorphic antibody genes encountered in a specific ethnic group's repertoire. RT-PCR was employed by Xoma to amplify the VH, Vk, and Vλ chains in 30 ethnically diverse healthy donors and tissues. There were 20 PBMC samples, eight bone marrow samples, one spleen sample, and one lymph node sample among the tissue sample. All Immunoglobulin (Ig) classes were included in the amplification strategy: IgM, IgG, IgA, IgE, and IgD. The libraries were displayed in Fab or scFv formats to assess potential differences in antibody selection based on display format. Both libraries generated a comparable number of distinctive antibodies with comparable affinity, indicating that the display format had no overall impact on the selection results. HAL9/10 also used blood samples from people of various ethnicities, including Caucasians, Africans, Indians, and Chinese. The libraries, one kappa (HAL10) and one lambda (HAL9), shared the same VH repertoire obtained from 98 donors but had different light chain repertoires. HAL9 contained all lambda subfamilies from the 98 donors, whereas HAL10 contained all kappa families from 54 donors with the notable exception of the IGKV7 pseudo-gene. In the amplification strategy, a reverse primer for V regions derived from IgMs was used, favoring the amplification of naive antibody genes, that is, those close to the germ-line configuration and thus with few or no somatic mutations in VH. A larger number of donors made the KNU-Fab library possible than the Xoma and HAL9/10 libraries. There had been 809 samples collected from 803 PBMC donors, two lymph node donors, two spleen donors, and two bone marrow donors. Thirty-three PBMC samples were obtained from healthy Korean human donors, whereas the 770 samples were obtained from commercial samples, most likely from donors of different nationalities. Because of antibody genes were amplified using forward and reverse primers that hybridized in the V regions, it was impossible to determine the Ig class, isotype, or whether the antibodies were close to the germline gene configuration or had a significant number of somatic mutations [39].

To summarize, various pools of human donors were used to construct the naive libraries. In some cases, such as the Xoma and KNU-Fab libraries, the authors cast as wide a net as possible to amplify all human genes, regardless of isotype or number of somatic mutations. In other cases, such as HAL9/10, the amplification strategy was restricted to germline gene sequences. Only a few antibody genes represented the majority of the studied sequences in all of the libraries, which is consistent with the patterns observed in human antibody repertoires. The HCDR3 length distribution was also consistent across all libraries, mirroring the Gaussian distribution seen in human antibodies [39].

4.2 Synthetic libraries

Synthetic repertoires are intended to maximize the functionality of antibody libraries by using well-expressed and developable scaffolds, targeting positions for diversification that do not disrupt V region folding, and selecting types and frequencies of amino acids that facilitate the selection of diverse binders to any given target. The first synthetic libraries, HuCAL and HuCAL GOLD, were developed by MorphoSys in the late 1990s. HuCAL PLATINUM, like its predecessor HuCAL GOLD, was built with seven VH and seven VL master scaffolds, which when combined yielded 49 antibody sub-libraries. These scaffolds were created using consensus sequences from the human repertoire's IGV gene families. The scaffold sequences were optimized for *E. coli* expression and displayed on the phage [40]. One of the primary distinctions between GOLD and PLATINUM is that the latter contains newly designed HCDR3 sequences. The new strategy was developed after a thorough examination of the amino acid usage per position of HCDR3 sequences with varying loop lengths. Instead of a relative uniform amino acid distribution for all HCDR3 of diverse lengths, different amino acid frequencies were used to design the synthetic HCDR3 fragments based on the specific length-dependent amino acid use frequencies. Furthermore, PLATINUM used loop lengths ranging from 4 to 23 amino acids, covering more than 95% of the naturally occurring HCDR3 lengths. Furthermore, potential N-glycosylation sites generated by the NXT/S pattern in HuCAL and

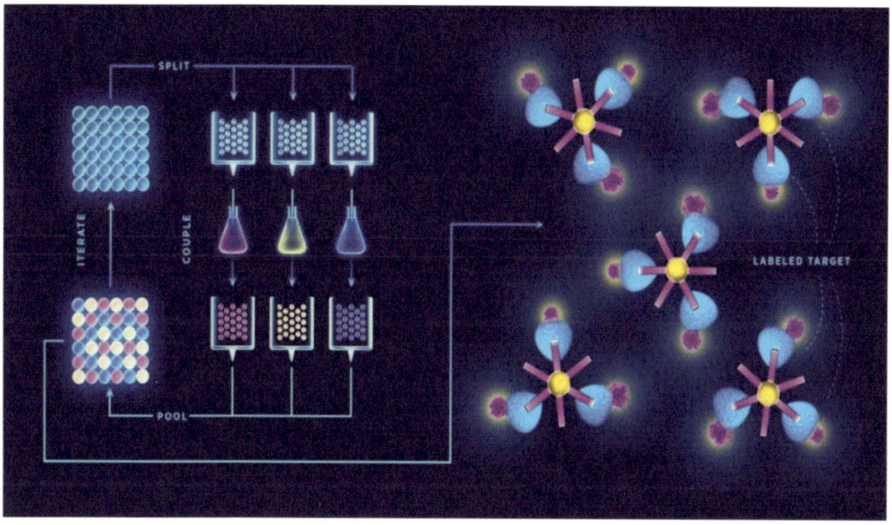

Figure 3.
Synthesis of combinatorial peptide libraries by the "split-and-mix" method. Carrier beads are split into aliquots (only 3 are shown for clarity) for coupling individual amino acid residues, pooled and the process is iterated until the desired length of peptides is achieved. Source: Bozovičar and Bratkovič, [42]; with permission.

GOLD (where N is asparagine, X is any amino acid, and S/T is serine/threonine) were removed from PLATINUM. A side-by-side comparison of HuCAL GOLD and PLATINUM revealed that the latter generated approximately 4-fold more unique and higher affinity antibodies as a result of these changes [41] (**Figure 3**).

5. Phage display libraries as an antibody discovery platforms

Antibody phage display is a versatile *in vitro* selection technology that can be employed to identify high-affinity antibodies specific to a wide range of antigens. However, specificity and affinity are not the only factors that contribute to the success of therapeutic antibodies. Other antibody quality characteristics, such as solubility, viscosity, expression yield, and thermal and long-term stability, are critical to mAb lead candidate success in biomanufacturing and clinical trials. Antibody biophysical properties are heavily influenced by their amino acid sequences. Because of their high immunogenicity, physicochemical instability, self-association, high viscosity, poly-specificity, short half-life, and poor expression, some mAbs may have poor developability profiles. Low solubility, for instance, can create issues during biomanufacturing and affect mAb potency, bioavailability, and immunogenicity. High thermal stability is required to maintain structural and functional integrity, as well as intrinsic properties, at varying temperatures. Furthermore, because of immunogenicity concerns, aggregation is one of the major challenges limiting the advancement of therapeutic mAb [43–45]. Despite the fact that antibody phage display has several advantages, including the ability to isolate antibodies against toxic or non-immunogenic antigens and generate conformation-specific antibodies, the vast majority of approved therapeutic antibodies are derived from immunized mice technologies. This is due to the immune system's purification process, which allows mammal-derived antibodies to still have better biophysical properties than phage display antibodies. Jain et al. examined the biophysical properties of 46 FDA-approved therapeutic antibodies and 89 in advanced clinical trials, which agreed with this. They discovered that antibodies discovered directly by phage display or engineered at some point by phage bio-panning have significantly lower developability risks than immunized mouse antibodies. Because they contain a larger percentage of aliphatic residues in their CDRs, phage display-derived therapeutic antibodies have higher self-interaction and poly-reactivity than non-phage-derived antibodies [45, 46].

Furthermore, because antibodies from phage display libraries are not glycosylated because they are produced in *E. coli*, some candidates may be negatively impacted when glycosylation occurs during mammalian cell expression; their binding, biodistribution, or pharmacokinetics may be impacted. As a result, it could be advantageous to be using eukaryotic display platforms such as yeast and mammalian display. Yeast and mammalian antibody libraries can be built to display full-length antibodies as well as antibody fragments such as scFvs and fragment antigen-binding region (Fabs), allowing the isolation of high-affinity antibodies with distinct biological properties. Parthiban *et al.*, for example, used CRISPR/Cas9 or transcription activator-like effector nucleases to create mammalian libraries that display approximately 10 million clones in IgG-format on the surface of HEK293 cells (TALENs). These libraries can act as a quality filter for various aspects of antibody development, as well as provide very early insight into issues such as aggregation and cross-reactivity. Each display system has advantages and disadvantages, but determining those is beyond the scope of this review, which focuses on phage display, the most frequently used type of antibody display. As a result, constructing phage libraries that enable the isolation of highly specific and diverse mAbs with high affinity against diverse antigens and optimal developability potential is critical [47, 48].

Biopharmaceutical companies are currently implementing extensive developability assessments in industrial pipelines to determine the biochemical and biophysical features of antibody candidates in order to aid in the identification of candidates with more favorable biophysical properties and to avoid difficulties during the downstream process. *In silico* platforms, such as the Therapeutic Antibody Profiler (TAP) tool, for example, are used as a warning system to predict mAbs with poor developability profiles by identifying anomalous values when compared to therapeutic mAbs in clinical stages. Indeed, algorithmic predictions anticipated that features within the variable regions of mAbs, such as total CDRs length, high hydrophobicity of VH and VL chains, lack of net charge symmetry, and/or the presence of patches of positive and negative charges, would indeed be significant considerations in mAb developability profiles. The likelihood of isolating high affinity, more diverse mAbs that specifically bind random epitopes increases significantly when bio-panning campaigns use larger antibody libraries. The diversity of a library is measured by how many functional antibody fragments can recognize as many different antigens as possible. The bacterial transformation step during library construction, on the other hand, is a major practical bottleneck that prevents the library from growing beyond 1011 antibody variants, even after optimization and numerous electroporation steps [49, 50].

Antibody phage display libraries should theoretically at least not only be large and diverse but should also demonstrate antibody variants as functional fragments. Nucleotide sequence issues, such as the presence of stop codons or the addition/deletion of nucleotides, can emerge during the library construction process. These issues may inhibit the production of functional pIII-antibody fusions or change the reading frame of the antibody gene sequence, affecting their biophysical properties. Some in-frame antibody genes may also have low phagemid expression levels or produce aggregated, misfolded, or toxic antibody fragments to *E. coli*. However, when compared to other variants or phages that do not display any fusion protein, such variants are usually displayed in a lower percentage [51, 52].

6. Bio-panning for target-specific antibodies

When purified antigens are available, they can be immobilized on solid surfaces such as nitrocellulose membranes, polystyrene tubes or plates, magnetic beads, or column matrices as well as presented to a phage antibody library. Nonspecific phage binding can be prevented by using blocking agents, such as bovine serum albumin (BSA), milk, or casein. After the phage library has already been exposed to the immobilized antigens, unbound phages are typically washed away (**Figure 4**). This step is critical for removing nonspecific binders and allowing some control over binding properties by adjusting the wash buffer and washing stringency. Long wash times, for example, can be used to ensure that only clones with slow dissociation rates are chosen. Wash buffers typically contain detergents, but they can also be tailored to account for variables such as pH and salt concentration. With each round of bio-panning, the washing steps are gradually increased to increase the stringency in order to isolate higher affinity phage clones [54, 55]. Different elution conditions, such as pH change, proteolytic cleavage, or competition with free antigens, have been used to recover high-affinity phage antibodies from immobilized antigens. Acidic buffers, such as glycine or citric acid, or alkaline triethylamine (TEA), can be used for pH elution. To avoid phage degradation and maintain infectivity, the pH of eluted phage

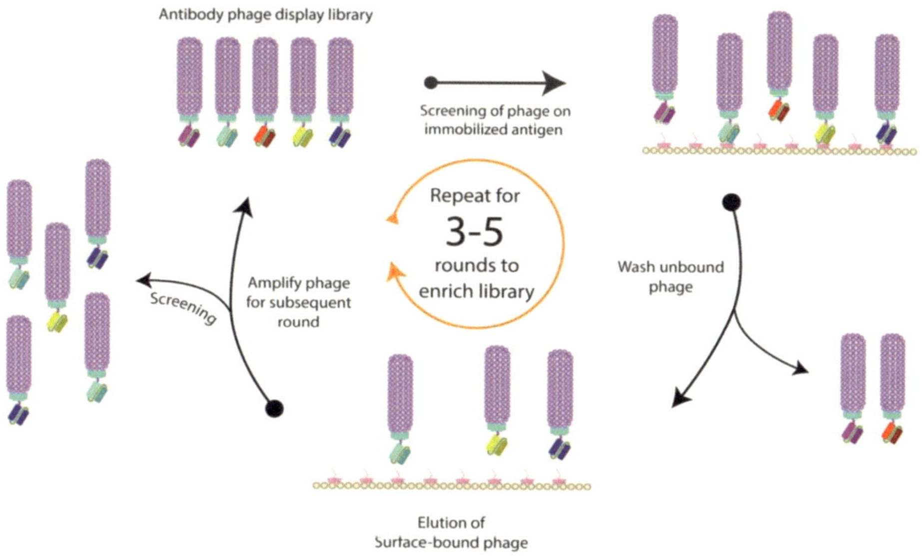

Figure 4.
Schematic representation of phage bio-panning. This is the basic method for sequential affinity screening of the phage display libraries for specific binding phage from a large excess of nonbinding clones is often referred to as "panning or bio-panning." Source: Alfaleh et al. [53]; with permission.

antibodies must be neutralized to around 8. To facilitate elution with proteases such as Genenase I or trypsin, some libraries include a cleavage site between the antibody and the pIII protein [56].

Following several rounds of bio-panning, the pool of phages isolated from each round is tested, typically using ELISA, to determine whether the polyclonal pool contains an enrichment of phage binders toward the specific antigen. Polyclonal ELISA begins while also immobilizing antigen on microtiter plates, followed by the incorporation of various dilutions of the phage pool from each round, and finally the identification of bound phage with an anti-M13 phage antibody. Individual clones with the highest enrichment level from the bio-panning round are then screened using ELISA to identify individual phage isolates with high specificity toward the antigen of interest. In a 96-well plate format, single colonies of cell glycerol stock from the most recent bio-panning round would be grown before the helper phage is decided to add to induce phage particle establishment. The positive clones derived from this experiment can then be analyzed using restriction fragment length polymorphism to determine the number of unique clones or sequencing to determine the CDRs for both the heavy and light chains. Once positive clones have been isolated, downstream applications will dictate how they are processed. A phage clone's scFv gene, for example, can be re-cloned into a bacterial expression vector for large-scale production, or it can be reformatted into a full mAb by inserting the variable regions into expression vectors containing the antibody constant regions. To increase the affinity and stability of the selected antibodies, *in vitro* affinity maturation can be performed using mutated libraries of lead phage clones [57, 58]. Although bio-panning of immobilized antigen on solid surfaces is effective, it is frequently hampered by a lack of purified protein and the possibility of the protein modifying confirmation when attached to solid surfaces. As a result, alternative approaches have been employed, such as in-solution bio-panning followed by affinity capture of biotin or calmodulin-binding peptide-tagged antigens [59].

Due to their complexity, some membrane proteins are poorly soluble in aqueous media and do not form properly during recombinant expression. When coated on immunotubes before bio-panning, they may form aggregates and lose their tertiary structures, resulting in antibody binders that recognize epitopes that are not naturally exposed. As a result, cell-based bio-panning is frequently used to keep membrane proteins in their native conformation. It can be used to retrieve antibodies specific for either known or unknown antigens on the cell surface and it can be used when the targeted antigen is not available in pure form. Selection of binders to a specific conformational state of a cell surface receptor is really possible by utilizing cell-based bio-panning strategies [60–62]. The most recent advances in next-generation sequencing (NGS) technologies, bioinformatics, and nanotechnology have greatly improved antibody discovery high-throughput screening. Raftery et al., for example, described a rapid selection of scFv-phage (PhageXpress) using electrohydrodynamic manipulation of a solution containing phage library particles in conjunction with Oxford Nanopore Technologies' MinION sequencer. They were able to identify 14 anti-dengue viruses nonstructural protein one scFv after a single round of bio-panning and in 2 days as opposed to several weeks using traditional bio-panning. Adopting similar approaches will significantly reduce the amount of time and laborious lab work required to discover putative antibodies, which are major impediments in the traditional bio-panning method, and will aid in the development of therapeutic monoclonal antibodies during emerging infectious outbreaks [63–65].

DOI: http://dx.doi.org/10.5772/intechopen.110729

Author details

Abebe Zemene
University of Gondar, Gondar, Ethiopia

*Address all correspondence to: abebeabuhayy@gmail.com

References

[1] Paria D, Venugopalan PL. Controlling evolution: The Nobel Prize in chemistry 2018. IEEE Pulse. 2019;**10**:12-16. DOI: 10.1109/MPULS.2019.2922566

[2] Korendovych IV. Rational and semirational protein design. Methods Molecules O Biologico. 2018;**1685**:15-23. DOI: 10.1007/978-1-4939-7366-8-2

[3] Lu RM, Hwang YC, Liu IJ, Lee CC, Tsai HZ, Li HJ, et al. Development of therapeutic antibodies for the treatment of diseases. Journal of Biomedical Science. 2020;**27**:1

[4] Ledsgaard L, Kilstrup M, Karatt-Vellatt A, McCafferty J, Laustsen AH. Basics of antibody phage display technology. Toxins. 2018;**10**:236

[5] Antonara S, Chafel RM, LaFrance M, Coburn J. Borrelia burgdorferi adhesins identified using in vivo phage display. Molecular Microbiology. 2007;**66**:262-276

[6] Ryvkin A, Ashkenazy H, Weiss-Ottolenghi Y, Piller C, Pupko T, Gershoni JM. Phage display peptide libraries: Deviations from randomness and correctives. Nucleic Acids Research. 2018;**2018**(18):46

[7] Kong X-D, Carle V, Diaz-Perlas C, Butler K, Heinis C. Generation of a large peptide phage display library b self-ligation of whole plasmid PCR product. Chemistry & Biology. 2020;**2020**(15):2907-2915

[8] Weber M, Bujak E, Putelli A, Villa A, Matasci M, Gualandi L, et al. A highly functional synthetic phage display library containing over 40 billion human antibody clones. PLoS One. 2014;**9**:e100000

[9] Nelson RS, Valadon P. A universal phage display system for the seamless construction of Fab libraries. Journal of Immunological Methods. 2017;**450**:41-49

[10] Zhoa Q, Buhr D, Gunter C, Frenette J, Ferguson M, Sanford E, et al. Rational library design by functional CDR resampling. New Biotechnology. 2018;**45**:89-97

[11] Tiller T, Schuster I, Deppe D, Siegers K, Strohner R, Herrmann T, et al. A fully synthetic human Fab antibody library based on fixed VH/VL framework pairings with favourable biophysical properties. MAbs. 2013;**5**:445-470

[12] Chan CE, Lim AP, MacAry PA, Hanson BJ. The role of phage display in therapeutic antibody discovery. International Immunology. 2014;**26**:649-657. DOI: 10.1093/intimm/dxu082

[13] Steinwand M, Droste P, Frenzel A, Hust M, Dubel S, Schirrmann T. The influence of antibody fragment format on phage display based affinity maturation of IgG. MAbs. 2014;**6**:204-218. DOI: 10.4161/mabs.27227

[14] Schofield DJ, Pope AR, Clementel V, Buckell J, Chapple S, Clarke KF, et al. Application of phage display to high throughput antibody generation and characterization. Genome Biology. 2007;**8**:R254. DOI: 10.1186/gb-2007-8-11-r254

[15] Turunen L, Takkinen K, Soderlund H, Pulli T. Automated panning and screening procedure on microplates for antibody generation from phage display libraries. Journal of Biomolecular Screening. 2009;**14**:282-293. DOI: 10.1177/1087057108330113

[16] Caceres DH, Forsberg K, Welsh RM, Sexton DJ, Lockhart SR,

Jackson BR, et al. Candida auris: A review of recommendations for detection and control in healthcare settings. Journal of Fungi (Basel). 2019;**5**:111

[17] El-Manzalawy Y, Dobbs D, Honavar VG. In silico prediction of linear B-cell epitopes on proteins. Methods in Molecular Biology. 2017;**1484**:255-264

[18] Sutandy FX, Qian J, Chen CS, Zhu H. Overview of protein microarrays. Current Protocols in Protein Science. 2013;**27**:1

[19] Howick VM, Russell AJC, Andrews T, Heaton H, Reid AJ, Natarajan K, et al. The malaria cell atlas: Single parasite transcriptomes across the complete Plasmodium life cycle. Science. 2019;**365**:eaaw2619

[20] Batinovic S, Wassef F, Knowler SA, Rice DTF, Stanton CR, Rose J, et al. Bacteriophages in natural and artificial environments. Pathogens. 2019;**8**:100. DOI: 10.3390/pathogens8030100

[21] Vikram A, Tokman JI, Woolston J, Sulakvelidze A. Phage biocontrol improves food safety by significantly reducing the level and prevalence of Escherichia coli O157:H7 in various foods. Journal of Food Protection. 2021;**83**:668-676. DOI: 10.4315/0362-028X.JFP-19-433

[22] Dunne M, Rupf B, Tala M, Qabrati X, Ernst P, Shen Y, et al. Reprogramming bacteriophage host range through structure-guided Design of Chimeric Receptor Binding Proteins. Cell Reports. 2019;**29**:1336-1350. DOI: 10.1016/j.celrep.2019.09.062

[23] Hashida Y, Tanaka H, Zhou S, Kawakami S, Yamashita F, Murakami T, et al. Photothermal ablation of tumor cells using a single-walled carbon nanotube peptide composite. Journal of Controlled Release. 2014;**173**:59-66

[24] Straus SK, Bo HE. Filamentous bacteriophage proteins and assembly. In: Harris J, Bhella D, editors. Virus Protein and Nucleoprotein Complexes. Singapore: Springer; 2018. pp. 261-279. DOI: 10.1007/978-981-10-8456-0_12

[25] Zhang X, Zhang X, Gao H, Qing G. Phage display derived peptides for Alzheimer's disease therapy and diagnosis [Internet]. Vol. 12. Theranostics. Ivyspring International Publisher; 2022. pp. 2041-2062. DOI: 10.7150/thno.68636

[26] Hassapis KA, Stylianou DC, Kostrikis LG. Architectural insight into Inovirus-associated vectors (IAVs) and development of IAV-based vaccines inducing humoral and cellular responses: Implications in HIV-1 vaccines. Viruses. 2014;**6**:5047-5076. DOI: 10.3390/v6125047

[27] Prisco A, De Berardinis P. Filamentous bacteriophage Fd as an antigen delivery system in vaccination. International Journal of Molecular Sciences. 2012;**13**:5179-5194. DOI: 10.3390/ijms13045179

[28] Tao P, Zhu J, Mahalingam M, Batra H, Rao VB. Bacteriophage T4 nanoparticles for vaccine delivery against infectious diseases. Advanced Drug Delivery Reviews. 2018;**145**:57-72. DOI: 10.1016/j.addr.2018.06.025

[29] Deng X, Wang L, You X, Dai P, Zeng Y. Advances in the T7 phage display system (Review). Molecular Medicine Reports. 2018;**17**(1):714-720

[30] González-Cano P, Gamage LN, Marciniuk K, et al. Lambda display phage as a mucosal vaccine delivery vehicle for peptide antigens. Vaccine. 2017;**35**(52):7256-7263. DOI: 10.1016/j.vaccine.11.010

[31] Cui Z, Gorzelnik KV, Chang JY, et al. Structures of Qβ virions, virus like

particles, and the Qβ-MurA complex reveal internal coat proteins and the mechanism of host lysis. Proceedings of the National Academy Science USA. 2017;**114**(44):11697-11702. DOI: 10.1073/pnas.1707102114

[32] Bookstaver ML, Tsai SJ, Bromberg JS, Jewell CM. Improving vaccine and immunotherapy design using biomaterials. Trends in Immunology. 2018;**39**(2):135-150. DOI: 10.1016/j.it.2017.10.002

[33] Christoph M, Stanley JR. Antibody phage display: technique and applications. The Journal of Investigative Dermatology;**134**(2):e17

[34] Zadeh AS, Grässer A, Dinter H, Hermes M, Schindowski K. Efficient construction and effective screening of synthetic domain antibody libraries. Methods Protocol. 2019;**2**:17

[35] Chan SK, Lim TS. Immune human antibody libraries for infectious diseases. Pharmaceutical Biotechnology. 2017;**2017**:61-78

[36] Van der Woning B, De Boeck G, Blanchetot C, Bobkov V, Klarenbeek A, Saunders M, et al. DNA immunization combined with scFv phage display identifies antagonistic GCGR specific antibodies and reveals new epitopes on the small extracellular loops. mAbs. 2016;**8**:1126-1135

[37] Hoet RM, Cohen EH, Kent RB, Rookey K, Schoonbroodt S, Hogan S, et al. Generation of high-affinity human antibodies by combining donor-derived and synthetic complementarity-determining-region diversity. Nature Biotechnology. 2005;**23**:344-348

[38] Almagro JC, Pedraza-Escalona M, Arrieta HI, Pérez-Tapia SM. Phage Display Libraries for Antibody Therapeutic Discovery and Development [Internet]. Vol. 8. Antibodies: MDPI AG; 2019. p. 44. DOI: 10.3390/antib8030044

[39] Glanville J, Zhai W, Berka J, Telman D, Huerta G, Mehta GR, et al. Precise determination of the diversity of a combinatorial antibody library gives insight into the human immunoglobulin repertoire. Proceedings of the National Academy Science USA. 2009;**106**:20216-20221

[40] Rothe C, Urlinger S, Löhning C, Prassler J, Stark Y, Jäger U, et al. The human combinatorial antibody library HuCAL GOLD combines diversification of all six CDRs According to the Natural Immune System with a Novel Display Methods for Efficient Selection of High-Affinity Antibodies. Journal of Molecular Biology. 2008;**376**:1182-1200

[41] Prassler J, Thiel S, Pracht C, Polzer A, Peters S, Bauer M, et al. HuCAL PLATINUM, a Synthetic Fab Library optimized for sequence diversity and superior performance in mammalian expression systems. Journal of Molecular Biology. 2011;**413**:261-278

[42] Bozovičar K, Bratkovič T. Evolving a Peptide: Library Platforms and Diversification Strategies [Internet]. International Journal of Molecular Sciences. MDPI AG. 2019;**21**:215. DOI: 10.3390/ijms21010215

[43] Lecerf M, Kanyavuz A, Lacroix-Desmazes S, Dimitrov JD. Sequence features of variable region determining physicochemical properties and polyreactivity of therapeutic antibodies. Molecular Immunology. 2019;**112**:338-346. DOI: 10.1016/j.molimm.2019.06.012

[44] Sormanni P, Amery L, Ekizoglou S, Vendruscolo M, Popovic B. Rapid and accurate in silico solubility screening of

a monoclonal antibody library. Scientific Reports. 2017;**7**:8200. DOI: 10.1038/s41598-017-07800-w

[45] Jain T, Sun T, Durand S, Hall A, Houston NR, Nett JH, et al. Biophysical properties of the clinical stage antibody landscape. Proceedings of the National Academy Science USA. 2017;**114**:944-949. DOI: 10.1073/pnas.1616408114

[46] Kaleli NE, Karadag M, Kalyoncu S. Phage display derived therapeutic antibodies has enriched aliphatic content: insights for developability issues. Proteins. 2019;**87**:607-618. DOI: 10.1002/prot.25685

[47] Nguyen AW, Le KC, Maynard JA. Identification of high affinity HER2 binding antibodies using CHO Fab surface display. Protein Engineering, Design & Selection. 2018;**31**:91-101. DOI: 10.1093/protein/gzy004

[48] Wolf Pérez A-M, Sormanni P, Andersen JS, Sakhnini LI, Leon I, Bjelke JR, et al. In vitro and in silico assessment of the developability of a designed monoclonal antibody library. mAbs. 2019;**11**:388-400. DOI: 10.1080/19420862.2018.1556082

[49] Bauer J, Mathias S, Kube S, Otte K, Garidel P, Gamer M, et al. Rational optimization of a monoclonal antibody improves the aggregation propensity and enhances the CMC properties along the entire pharmaceutical process chain. mAbs. 2020;**12**:1787121. DOI: 10.1080/19420862.2020.1787121

[50] Raybould MIJ, Marks C, Krawczyk K, Taddese B, Nowak J, Lewis AP, et al. Five computational developability guidelines for therapeutic antibody profiling. Proceedings of the National Academy Science USA. 2019;**116**:4025-4030. DOI: 10.1073/pnas.1810576116

[51] Gupta A, Shrivastava N, Grover P, Singh A, Mathur K, Verma V, et al. A novel helper phage enabling construction of genome scale ORF-enriched phage display libraries. PLoS One. 2013;**8**:e75212. DOI: 10.1371/journal.pone.0075212

[52] Kulmala A, Huovinen T, Lamminmäki U. Effect of DNA sequence of Fab fragment on yield characteristics and cell growth of *E. coli*. Scientific Reports. 2017;**7**:3796. DOI: 10.1038/s41598-017-03957-6

[53] Alfaleh MA, Alsaab HO, Mahmoud AB, Alkayyal AA, Jones ML, Mahler SM, et al. Phage Display Derived Monoclonal Antibodies: From Bench to Bedside [Internet]. Frontiers in Immunology. Frontiers Media SA. 2020;**11**. DOI: 10.3389/fimmu.2020.01986

[54] Schirrmann T, Meyer T, Schütte M, Frenzel A, Hust M. Phage display for the generation of antibodies for proteome research, diagnostics and therapy. Molecules. 2011;**16**:412-426. DOI: 10.3390/molecules16010412

[55] Shen W, Li S, Park M-K, Zhang Z, Cheng Z, Petrenko VA, et al. Blocking agent optimization for nonspecific binding on phage based magnetoelastic biosensors. Journal of the Electrochemical Society. 2012;**159**:B818-B823. DOI: 10.1149/2.057210jes

[56] Haque A, Tonks NK. The use of phage display to generate conformation-sensor recombinant antibodies. Nature Protocols. 2012;**7**:2127-2143. DOI: 10.1038/nprot.2012.132

[57] Pande J, Szewczyk MM, Grover AK. Phage display: concept, innovations,

applications and future. Biotechnology Advances. 2010;**28**:849-858. DOI: 10.1016/j.biotechadv.2010.07.004

[58] Jones ML, Seldon T, Smede M, Linville A, Chin DY, Barnard R, et al. A method for rapid, ligation-independent reformatting of recombinant monoclonal antibodies. Journal of Immunological Methods. 2010;**354**:85-90. DOI: 10.1016/j.jim.2010.02.001

[59] Mukherjee S, Ura M, Hoey RJ, Kossiakoff AA. A new versatile immobilization tag based on the ultra-high affinity and reversibility of the calmodulin-calmodulin binding peptide interactions. Journal of Molecular Biology. 2015;**427**:2707-2725. DOI: 10.1016/j.jmb.2015.06.018

[60] Jones ML, Alfaleh MA, Kumble S, Zhang S, Osborne GW, Yeh M, et al. Targeting membrane proteins for antibody discovery using phage display. Scientific Reports. 2016;**6**:26240. DOI: 10.1038/srep26240

[61] Alfaleh AM, Jones LM, Howard BC, Mahler MS. Strategies for selecting membrane protein-specific antibodies using phage display with cell-based panning. Antibodies. 2017;**6**:10. DOI: 10.3390/antib6030010

[62] Alfaleh AM, Arora N, Yeh M, de Bakker JC, Howard BC, Macpherson P, et al. Canine CD117-specific antibodies with diverse binding properties isolated from a phage display library using cell-based biopanning. Antibodies. 2019;**8**:15. DOI: 10.3390/antib8010015

[63] Yoo DK, Lee SR, Jung Y, Han H, Lee HK, Han J, et al. Machine learning-guided prediction of antigen-reactive in silico clonotypes based on changes in clonal abundance through bio-panning. Biomolecules. 2020;**10**:421. DOI: 10.3390/biom10030421

[64] Raftery LJ, Howard CB, Grewal YS, Vaidyanathan R, Jones ML, Anderson W, et al. Retooling phage display with electrohydrodynamic nanomixing and nanopore sequencing. Lab on a Chip. 2019;**19**:4083-4092. DOI: 10.1039/C9LC00978G

[65] Kelley B. Developing therapeutic monoclonal antibodies at pandemic pace. Nature Biotechnology. 2020;**38**:540-545. DOI: 10.1038/s41587-020-0512-5

Chapter 4

Polymerase Chain Reaction: Applications in Gene and Cell Therapy Studies

Jacqueline Murphy, Kate Herr and Venkata Vepachedu

Abstract

The rapidly developing fields of gene and cell therapy allow us a platform to repair or replace defective genes or introduce a missing gene. AAV and lentivirus are common viral vectors used in gene therapy to deliver a DNA payload to a tissue of interest. Recently, self-replicating RNA-based vaccines and therapies are also becoming increasingly popular for gene therapy after the success of SARS-CoV-2 vaccines. Cell therapy is the transplantation of human cells without or with *ex vivo* modification utilizing CAR-T and stem cell technology. Because PCR allows us to detect transgenes with high sensitivity, we can leverage this technology to quantify the efficacy of a therapy and long-term expression *in vivo* using both qPCR and RT-qPCR, respectively. PCR provides information that is used to justify first in human dose, toxicological evaluations, efficacy through PK/PD relationships, monitor persistency and shedding as well as biomarker and gene expression quantitation. As evaluation of safety endpoints is critical to drug development, PCR is imperative in the field of clinical pharmacology discovery.

Keywords: biodistribution, qPCR, RT-qPCR, adeno-associated virus (AAV), CAR-T, transgene, probe, multiplex, matrix, digital PCR, shedding

1. Introduction to gene therapy

Countless human diseases are caused by a change or mutation to a single gene. Examples include the CFTR gene in Cystic Fibrosis and the Factor IX protein in hemophilia making these diseases strong candidates for gene therapy. In an effort to treat or cure a disease, gene therapy attempts to introduce a functional protein to the diseased tissue or cells using a viral vector. The most common vector used in gene therapies is AAV (adeno-associated virus), which is a naturally occurring non-pathogenic parvovirus with a single-stranded, linear DNA genome about 4.7 kilobases in size. Once the viral DNA of the wild-type AAV is replaced with the therapeutic DNA also known as the transgene or corrected gene of interest, it is now considered recombinant AAV (rAAV). The rep and cap genes of the AAV serotype are now provided in trans during the production of vector. The packaging limit for AAV is approximately 5 kb although innovations have been made to overcome this. rAAV-mediated gene delivery has proven to have highly efficient transduction as well as long-term gene expression which is what makes this vector a great delivery system for gene therapies.

AAV gene therapy routes of administration can vary by indication or study type. The capsids can be injected directly into the tissue of interest such as intravitreal for eye indications or intramuscularly or intravenously. The typical mechanism of action of an AAV gene therapy is receptor mediated. The capsid will bind to the cell membrane and become endocytosed forming an endosome containing the viral particle. The capsid will escape from the late endosome and translocate to the nucleus where the protein capsid will be uncoated, and the single-stranded genome will be released. The formation of double-stranded DNA will occur in the nucleus, and the rep genes will begin to be expressed allowing for the packaged transgene to replicate inside the cell. AAVs rely on cellular polymerases to replicate its genome and packaged DNA inside the nucleus of transduced cells. This is where qPCR will quantitate the amount of transgene that can be detected inside the cell.

To measure the efficacy of a gene therapy, a biodistribution study is done. This consists of designing a small or large animal study administering the gene therapy and harvesting a panel of tissues to analyze biodistribution of the transgene. Small dissections of tissues from the dosed animals are harvested and processed for qPCR analysis. qPCR (quantitative polymerase chain reaction) is used to quantify the amount of transgene DNA sequences that has been transduced into the tissue of interest as well as peripheral non-targeted tissues. Bioanalysis results are reported as copies of the target DNA per µg of host genomic DNA. qPCR probes are designed to bind and amplify the transgene sequence and a copy number is calculated from the cycle threshold by applying a standardization curve of known copy numbers.

2. Introduction to cell therapy

Cellular therapy is the use of live cells as the drug product in the treatment of disease. The cells or "living drugs" are transplanted either with or without post *ex vivo* genetic modification and include Chimeric Antigen Receptor (CAR) T-cell therapy, stem cells, Engineered T Cell Receptor (TCR) Therapy, tumor-infiltrating lymphocytes (TILs), and NK cells with the potential for more therapies in the future. T cells are often selected because of the important role they play in cell-mediated immunity and offer a targeted therapy as an alternative to chemotherapy and/or surgery [1]. To highlight how important cell therapy scientific advancements of being able to harness the body's natural immunity to fight disease have been, we note that they resulted in the Nobel prize in Medicine being awarded to James P. Allison and Tasuku Honjo in 2018 [2]. Their research, along with many others, on how the immune system can be used in defense of cancer, has opened to the doors for clinical trials that have resulted in treatments for individuals who had no other treatment options. The FDA has approved the use of multiple cell therapies including Kymriah®, Yescarta®, Tecartus®, Breyanzi®, Abecma®, Carvykti®, Zyneglo®, and Skysona®.

The CAR-T field specifically is rapidly expanding with a shift in focus from leukemia and lymphoma to solid tumors that are often difficult to target as well as from autologous treatment, where cells are taken from individual patients, to allogenic where donor cells will be used. The latter would allow for a more "off-the-shelf" treatment option, which would allow for earlier intervention of the disease (Parker Institute 2019). Advancement in the cell therapy also includes developing new modalities that are emerging alongside progress in current technology.

Therefore, monitoring the presence of cell therapies and their products is of high importance when utilizing them as a treatment option. Therefore, it is essential to have tools to monitor for the presence of the cellular drug product post-dose, for biodistribution, for potential shedding, and for transgene expression if applicable. These are used to help correlate dose versus efficacy, to evaluate potential toxicity, which would affect safety, for evaluation of off-target distribution and clearance from the body. They are also used to assess potential risks to the patient or community through shedding studies to evaluate the presence of the viral particles. However, as cell therapy treatment potential increases, adverse effects also need to be considered as cytokine release syndrome (CRS) as well as immune effector cell-associated neurotoxicity syndrome (ICANS) can potentially occur [3]. Due to this, it is important to note that immunogenicity is also an important assessment that should be paired with cellular kinetics for full evaluation.

Flow cytometry has historically been used to measure the cellular kinetics of cell therapies and is now often paired or sometimes replaced by quantitative PCR measurement. These two widely used methodologies offer detection strategies that are critical for evaluating the appropriate dose to patients through detecting the cells in relation to the efficacy and potential adverse reactions. A strong correlation has been established between these two assay types when comparing vector copy number by qPCR and CAR expression by flow cytometry. qPCR as a methodology provides measurements used to detect vector or transgene DNA or mRNA by establishing a standard curve for semi-quantitative evaluation where a linear regression is applied to the Ct values and the log concentration of each dilution. Digital PCR applies absolute quantitation and may also be utilized where applicable. For CAR-T, a critical window is 28 days post-dose to measure peak expansion and persistence but needs to be measured in long term as well. While flow cytometry is a reliable method to detect circulating CAR-T cells, it utilizes fresh cells, which can prove difficult if not processed on-site or quickly after. PCR provides readouts at very low levels due to high sensitivity that allow for measurement after the dose and expansion phase to monitor persistence, which expands the time points that we can screen for CAR-T cells [3].

3. Regulated work

Bioanalysis utilizing qPCR, ddPCR, dPCR falls under the supervision of regulatory agencies including the FDA and EMA. As countries approve medical treatments, they release regulatory guidance that is heavily relied on in drug development and clinical testing to determine how testing should occur. Pre-clinical and clinical studies pre-determine the collection and analysis methods that will be used in the evaluation of the efficacy and safety of the drug.

The FDA and the EMA have released both final and draft guidance on cell and gene therapy itself but there are few regulated to validate bioanalysis specifically at this time. There are also ICH guidelines with a focus on nonclinical biodistribution, virus and vector shedding, and oncolytic viruses. However, the validation of the PCR-specific assays still has many areas that are not in focus. The MIQE guidelines combined with published white papers are currently utilized to provide alignment and harmonization across the industry as white papers published by experts in the field also provide direction when conducting regulated work. As draft and final guidance are released for the validation of these PCR assays, harmonization can be achieved.

4. Modalities

Gene therapy is classified into two types based on the cells treated or transgene delivered to. They are somatic and germline gene therapy. Due to ethical reasons and safety/efficacy concerns, viral vectors are considered unacceptable for germline gene therapy (GGT). For GGT only Mitochondrial Replacement Therapy to replace mutated mitochondrial DNA from oocytes with normal mtDNA is permitted in UK [4]. Gene editing tools such as CRISP6-Cas9 are considered as alternatives for GGT for non-mitochondrial GGT [5].

Somatic gene therapy on the other hand allows the use of a wide variety of modalities for gene delivery. This includes both biological and non-biological methods. Among the biological methods viral, non-viral vectors and genome editing tools such as CRISPR-Cas9 are in use. Currently, the viral vectors are the most successful and widely used [6]. The most popular viral vectors in use for gene therapy are adeno-associated virus, adenovirus, lentivirus, retrovirus, oncolytic virus (Herpes Simplex Virus-1). A brief description on the application of these viral and non-viral vectors in gene and cell therapy programs will be discussed.

4.1 Viral vectors

Typically, a viral vector is defined by three components as follows: (1) the transgene; (2) the envelope or capsid that harbors the payload, which also defines antigen recognition and tropism, and (3) the "regulatory cassette" that controls transgene expression with or without integration.

4.2 Adenovirus vectors (Ad vectors)

These are the first viral vectors used in gene therapy [7]. Ad is a non-enveloped virus, with icosahedral protein capsid harboring a 26- to 45-kb linear, double-stranded DNA genome comprising five each of early phase (E1A, E1B, E2, E3, and E4) and late phase genes (L1–L5). These genes are flanked by hairpin-like inverted terminal repeats (ITRs) ranging from 30 to 371 bp in length. E1A is essential for the transcription of other viral genes (E1B, E2-E4), and late phase genes are responsible for viral DNA synthesis and other roles. The Ad viral capsid is composed of hexon, penton, fiber and capsid protein precursors, and viral core proteins. There are about 100 genotypes of Ad viruses. First-generation Ad vectors were generated by replacing the E1A/E1B region with transgene cassettes upto ~4.5 kb. Second-generation Ad vectors were developed by deleting E2A, E2B, or E4 genes that allowed transgene cassettes upto 10.5 kb size. First- and second-generation Ad vectors require engineered cell lines for the deleted genes. Third-generation Ad vectors were developed by deleting all viral sequences except ITRs and the packaging signal (psi). These are called, "helper dependent" and "high capacity" (HCAds) as they can carry up to ~36 kb of the cassette. HCAds also have low immunogenicity [8, 9]. Recently, a PCR-based method termed as Restriction Assembly (RA) has been developed. This method uses long restriction-digested fragments and short PCR products for DNA assembly. This was used successfully to construct E1 or E1/E3 deleted vectors based on simian adenovirus 1 (SAdV-1) [10]. In addition, PCR-based directed evolution and capsid engineering are found to be useful to produce adenovirus vectors of the desired properties [11]. For example, novel chimeric rAd5 vectors were constructed replacing the seven short hypervariable regions (HVRs) on the surface of the Ad5 hexon protein with the corresponding HVRs from the rare adenovirus serotype Ad48. These vectors lack pre-existing anti-vector immunity against the key hexon surface protein epitopes [12]. Conditionally replicative Ad viruses were constructed (CRAd or rAd) by cloning under tumor-specific promoters to make oncolytic rAd viruses [13].

4.3 AAV vectors

Adeno-associated viruses are currently the leading platform for gene/cell therapy programs. Its genome is ~4.7 kb long single-stranded DNA. For essential genes needed for replication and gene expression, it depends on adenovirus or another virus, which can provide the same support. These functions are provided by the Ad E1, E2a, E4, and VA RNA genes. AAV genome has four ORFs, a) *rep* encodes the four replication genes, b) *cap* encodes three viral capsid proteins, c) the third and fourth are nested sub-genomic mRNAs—Assembly Activating Protein (AAP) involved in the shuttling of capsid monomers to the nucleolus for capsid assembly and Membrane Associated Accessory Protein (MAAP) the function of which is not yet clear [14]. The 4.7-kb genome is flanked by 145 base ITRs on both ends of the genome. The ITRs serve as self-priming structures for replication and provide the signal for Rep-mediated packaging.

Several serotypes of AAVs with varied tropism are identified and used in pre-clinical and clinical studies. The tropism depends on the capsid type [15]. So capsid modification enables the generation of recombinant vectors with tailored properties and tropism. Initial attempts to engineer the capsid and redirecting to the desired tissue were performed by PCR amplification of the capsid sequences of choice and grafting them to the selected AAV serotype genome. DNA shuffling, site-directed mutagenesis, and error-prone PCR are the PCR-based methods utilized in the rational and directed evolution of AAV capsids [16, 17]. For example, mutants were isolated by error-prone PCR and staggered extension protocol of diverse libraries of AAV5 capsids that have enhanced transduction in human hepatocytes and favorable serore-activities [18]. A robust 2-step PCR method for DNA shuffling of selected serotypes of AAVs to generate capsids of desired immunogenicity levels and tropisms has been developed [19]. For example, capsid DNA from AAV serotypes 1–6, 8, and 9 shuffled and recombined to create a library of chimeric AAVs. From these, two mutants selectively crossed the seizure-compromised BBB and transduced cells in mice [20].

4.4 Lentiviral vectors

Retroviruses are spherical, enveloped with two sense-stranded RNA of ~9.2 kb size. The RNA strands are bound by nucleocapsid proteins. Lentiviruses belong to the HIV-1 type of retroviruses [8]. It has three core genes, namely *gag* responsible for encoding the structural proteins, *pol* for reverse transcriptase, protease, and integrase, and *env* for virus envelope glycoprotein [21]. The genome has regulatory genes (*tat* for activating viral transcription and *rev*—for splicing and export of transcripts) as well as four accessory genes (*vif, vpr, vpu*, and *nef*). Accessory proteins are not essential for replication. Long terminal repeats are present on two ends. Both the 5′ and 3′ LTRs in wild-type HIV-1 is composed of U3, R, and U5 sequences. The "psi" just after LTR is important for signaling and genome encapsidation.

Lentiviral vectors integrate into host genome with a packaging capacity of ~9 kb. Lentiviral vector systems are improved over the year for safety and lower immunogenicity. They were more successful in *ex vivo* applications generating CART cells for cancer immunotherapy. The third generation of lentiviral vector carrying the gene of interest contains a promoter (CMV/RSV), part of the 3′U3 LTR. The psi (ψ) packaging signal is followed by the rev response element (RRE) [18]. A novel lentiviral vector (LTR1) described as the fourth-generation vector was designed with a unique genomic structure to prevent transfer of HIV-1 packaging sequences to patient cells. This reduces the total HIV-1 content to ~4.8% of the wild-type genome (**Figure 1**) [22]. Helper plasmid constructs provide other necessary genes for capsid formation, which do not replicate.

4.5 Chimeric antigen receptor T-(thymus) lymphocytes (CAR-T)

CARs are engineered synthetic receptors that function to redirect T cells, to recognize and eliminate cells expressing a specific target antigen. Chimeric antigen receptor (CAR) T-cell therapy is a major step in personalized cancer treatment. In CAR-T cell therapy, a patient's own T cells (allogenic) are engineered to express a synthetic receptor that binds a tumor antigen and injected back into the patient to attack and kill the cancer cells. Viral vectors are used to encode CARs; the retroviruses integrate their DNA in the genome of the obtained T cells for long-term expression. These viral vectors include retroviruses, lentivirus, adenovirus, and adeno-associated virus [23]. Over the time the CAR-T cell therapy has evolved from first generation to fifth generation. In the first generation, only immunoreceptor tyrosine-based activation motif (ITAM)—(CD3ζ) motifs as the T cell stimulatory molecule within the intracellular domain was present. The second- and third-generation CAR-T cells had one and two costimulatory molecules, respectively. The fourth generation was designed with one costimulatory molecule paired with cytokine expressors (e.g., IL-12) under the control of NFAT transcription factor, hence referred to as T cell redirected for universal cytokine mediated killing (TRUCKs). The fifth generation was also like fourth generation plus intracellular domains of cytokine receptors (e.g., IL-2Rβ) to

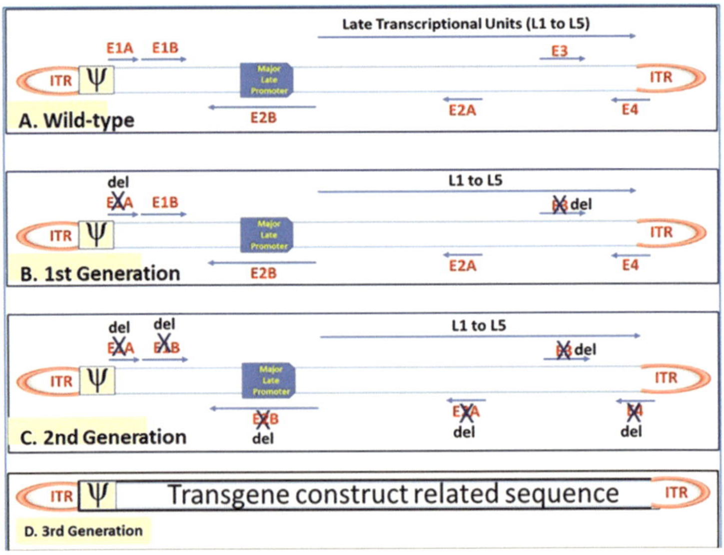

Figure 1.
Adenovirus vector generations. ITR = inverted terminal repeats, ψ = packaging signal, major late promoter (MLP), E1A, E1B, E2A, E2B, E3 and E4 – Early transcription units, L1-L5 – Late transcription units.

activate JAK and STAT3/5, stimulate cell proliferation, and enhance its persistence [24, 25]. The most recently FDA approved CAR T-cell therapy, in February 2022, is ciltacabtagene autoleucel (CARVYKTI™) from Janssen Therapeutics Inc. This drug was approved for the treatment of r/r multiple myeloma. It is a genetically modified autologous CAR T-cell therapy directed by B-cell maturation antigen (BCMA).

4.6 Stem cells

Any treatment that involves the use of viable human stem cells including embryonic stem cells (ESCs), iPSCs (induced pluripotent stem cells), and adult stem cells for autologous and allogeneic therapies is considered as stem cell therapy [26, 27]. Application of stem cells for therapeutic purposes could be achieved after the production of iPSCs *via* the transient over-expression of four transcription factors: OCT4, SOX2, KLF4, and MYC in fully differentiated somatic fibroblast cells [28]. Stem cells from different tissues are produced and applied for the treatment of diseases or impairments related to those organs with different rates of success. Integrating vectors effectively achieve long-term expression of a transgene in the treatment of monogenetic hematopoietic disorders. Accurate vector copy number (VCN) estimation has become increasingly important in stem cell therapies also. Application of both qPCR and dPCR methods was used in the vector copy number analysis of transduced stem cells [29].

4.7 Self-amplifying RNA

Self-amplifying RNA (saRNA) replicon is a type of RNA genome that is unique to certain families of viruses. This is due to the presence of RNA-dependent RNA Polymerase (RdRP) due to which the genomic RNA can make multiple copies of it finally resulting in the production of proteins, from just a single RNA template [30]. Some of the viruses used in the production of saRNA-dependent vaccines are Venezuelan Equine Encephalitis (VEE), Semliki Forest Virus (SFV), and Recombinant Sindbis/Venezuelan Equine Encephalitis Virus (VEE/SINV). A typical saRNA-based transgene construct contains from 5'end a) 5'CSE (Conserved Sequence Element) with 7-methylguanine, b) RdRP complex cassette from an saRNA virus (Alpha virus), c) a transgene/immunogen of interest, d) 3'CSE with polyA tail.

It is also possible to have more than one immunogen in one construct separated by self-splicing peptide sequences such as 2A or internal ribosome entry (IRE) site. However, in the clinical studies involving saRNAs, only RT-qPCR methods need to be developed for PK/biodistribution studies. saRNA constructs are usually delivered *via* lipid nanoparticles. This requires the formulation of the payloads after construction. It is necessary to monitor the toxicity due to lipid also in these studies (**Figure 2**).

4.8 RISC (RNA-induced silencing complex)

In diverse group of organisms, dsRNAs were shown to interfere with gene expression in a sequence-specific fashion in cultured cells [31]. This was first identified in plants, as "post-transcriptional gene silencing" (PTGS). RNA interference (RNAi) is a nucleic acid-based immune defense against viruses, transgenes and transposons. dsRNAs present in a cell are converted either through processing or replication, into small specificity determinants of discrete size in a manner analogous to antigen processing. Small interfering RNA (siRNA) and microRNA (miRNA) are small RNAs of

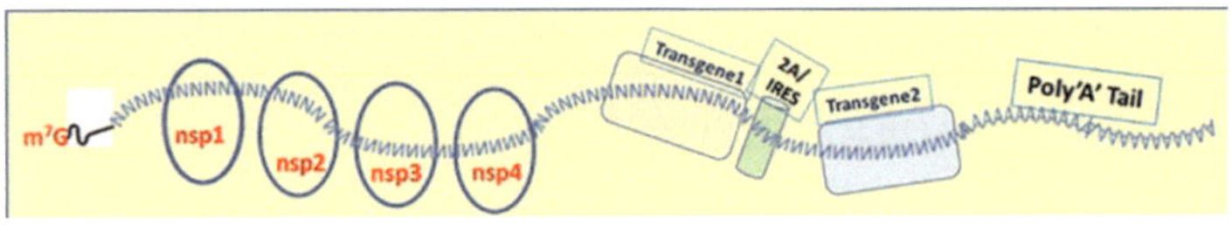

Figure 2.
Basic structure of a typical saRNA construct. nsp1-nsp4 form RNA-dependent RNA. Polymerase from alpha virus, required for self-replication of RNA. Transgenes (immunogens) are the genes of interest separated by either 2A or IRES sequences. 2A self-cleaves after translation at C-terminal separating the two transgenes. Internal ribosome entry site (IRES) permits translation of second transgene by a different mechanism. 2A/IRES are not required if there is only one transgene. Finally, there will be a 7-methyl guanosine at 5'end and poly a tail at 3' end.

18–25 nucleotides (nt) in length that play important roles in regulating gene expression. They are incorporated into an RNA-induced silencing complex (RISC) and serve as guides for silencing their corresponding target mRNAs based on complementary base-pairing [32]. The pathway of a typical eukaryotic gene silencing goes as follows:

a. Long dsRNAs are cleaved by Dicer, an RNAse III like enzyme, into small RNA duplexes of 21 to 23 nucleotides, called siRNA.

b. The Dicer-siRNA complex interacts with a cellular TAR (Transactive Response) RNA-binding protein (TRBP).

c. Another core protein of RISC complex is Argonaute 2 (Ago2). The Dicer-SiRNA complex binds with the Ago2-associated RISC complex. Some other components need to be identified.

d. After base pairing of the single-stranded guide RNA with the target mRNA, the PIWI (P-element Induced Wimpy) domain of the Ago2 protein is brought into proximity and cleaves the target mRNA resulting in a post-transcriptional gene silencing.

One major difference between siRNAs and miRNAs is that siRNAs are specific to a single mRNA target, whereas the miRNAs bind to multiple targets [33]. Therapeutic approaches based on siRNA involve the delivery of a synthetic siRNA into the target tissue/cells to elicit RNA interference (RNAi), thereby inhibiting the expression of a specific messenger RNA (mRNA) to produce a gene silencing effect. In contrast, miRNA-based therapeutics comprise two approaches: miRNA inhibition and miRNA replacement. Due to very short length, siRNA quantification by regular RT-qPCR is not feasible. For this reason, alternative PCR-based methods such as primer extension (PE), the invader assay, stem-loop RT-qPCR, ligation assay, and crook hairpins are developed and used. A modified form of competitive PCR [34] showed improved results with synthetic siRNA molecules.

5. PCR sample prep

To process tissue matrices for PCR amplification, the DNA and RNA need to be extracted and purified from the cells. Template quality is one of the biggest issues facing industry for bioanalysis. Any contaminant in the sample can cause inhibition and lead to biological non-reproducibility in the assay. There are many variations of nucleic acid extraction protocols that are used within the field. Many vendors provide kits for genomic DNA extraction or total RNA extraction, which are well characterized and used widely for optimal purity. Cells or tissues require pre-processing. The tissues need to be pulverized by mechanical processes such as a metal bead or grinding in liquid nitrogen to homogenize the sample. Proteinase K is an enzyme that can be used to digest peptide bonds to break down proteins that may contaminate your sample. RNase A can be used in DNA purifications as well to rid the sample of any RNA contamination. Once the pre-processing steps are completed, a detergent can be used to lyse the cells. Commercially available kits will call this a lysis buffer. The samples will quickly become viscous once the nucleic acid has broken out of the cells. The DNA will need to be pulled out of this solution in a binding step. It is most common to use a positively charged membrane or bead to bind to the negatively charged DNA molecules. Silica membranes with high concentrations of salt are also very common methods of binding genomic DNA so washing steps can be carried out without the risk of losing DNA. While the DNA is bound, ethanol is used to wash away all of the impurities from the cell lysate while still maintaining conditions to keep the binding strong. Contaminants and enzyme inhibitors should be washed away during these steps to ensure the sample is ready to be used in PCR. Water or TE can be added to elute the DNA off of the bound matrix into a purified sample ready for PCR.

Purifying RNA has the same overall procedure with a few steps catered toward the binding of RNA structure. Phenol-chloroform extractions were considered the gold standard extraction method for many years and are still used today; however, commercially available kits make the process much more high throughput and less risk of contamination. A DNase step would be done after purification of RNA to remove any DNA contamination. This ensures that in a RT-qPCR, all of the DNA that is amplified is from the cDNA synthesis step and not from any genomic DNA contamination from an impure RNA extraction. This is the reason to include a control sample in all RT-qPCR reactions, which excludes the reverse transcriptase in the reaction. If the sample shows any amplification, then it is amplifying genomic DNA contamination. Without the addition of the reverse transcriptase enzyme, cDNA synthesis cannot occur.

6. Standard curve

Quantitative PCR is a relative method, which means to assign absolute quantification to sample, an accurate and reproducible standard curve is required. It is acceptable to use a surrogate for the DNA being quantified in the experiment such as a DNA plasmid containing the sequence of interest. For gene therapy experiments, it is common to use the cis-plasmid used in AAV transfection, which contains the transgene sequence along with the entire sequence to be packaged into the AAV capsid for testing. Linearization of the circular plasmid by restriction digest is required so that the standard DNA most closely mimics that of the sample DNA.

Calculating the DNA concentration of the standard is essential for assay accuracy. Once the DNA concentration is measured using a spectrophotometer-based method such as a GloMax®, a copy number must be calculated in order to generate the standard curve. The equation below is used to convert nanograms per microliter (ng/uL), which is provided by the concentration instrumentation, into copies per microliter using the size of the DNA plasmid in base pairs, the molecular weight of a base pair of DNA (650 Da), and Avogadro's number.

$$DNA\left(\frac{copies}{\mu L}\right)=\frac{DNA\left(\frac{ng}{\mu L}\right)\times 10^{-9}\frac{g}{ng}\times 6.022\times 10^{23}\frac{molecules}{mole}}{N(bp)\times 650\frac{g}{mole\ of\ bp}}$$

This equation can also be used for calculating copies/uL of RNA; however, the average mass of 1 base pair of single stranded RNA (330 g/mol) must be substituted into the equation. The standard curve is a serial dilution that must cover all expected copy numbers of the unknown samples. If a sample is amplified at a Ct lower than the highest point on the standard curve, the sample will have to be diluted to fall within the curve to have a copy number assigned or a new standard curve generated for samples with higher copy numbers. The standard curve should be run in triplicate in the qPCR so that.

7. Probe/primer design for gene therapy studies/inhibition

7.1 AAV titering

PCR is a tool that can be used to both titer viruses and quantitate DNA delivered to cells by AAV vector. AAV titer is one of the limitations in the field of gene therapy. The maximum concentration that can be formulated from a transfection method is about 5E13 genome copies per mL. Human studies require high doses of gene therapies to see efficacious results so routes of administration as well as capsid engineering for more effective delivery methods is a means of overcoming this constraint. Once AAV virus particles are harvested from a transfection in HEK293 cell culture, qPCR will quantitate the amount of AAV in the purified formulation. There is typically a thorough purification process done before the sample is ready for titering, which can include ion-exchange columns, ultracentrifugation, filtering, DNase treatments to get rid of any unpackaged plasmid DNA as well as a proteinase K treatment to open the capsid and release the AAV DNA genome as a template for qPCR. A probe can be designed in any of the AAV genome that persists after packaging. This is commonly the polyA or the ITRs as this is highly conserved between serotypes so a single probe set can be used for titering many viruses.

7.2 Biodistribution

The Food and Drug Administration (FDA) recommends designing qPCR probes to amplify transgene DNA rather than other AAV genome sequences. This is to quantitate the therapeutic DNA directly rather than quantifying AAV genes such as ITRs, rep, or cap gene, which may not directly correlate with the transgene copy number due to impure capsid population. During the transfection process where AAV capsids are assembled and transgene DNA is packaged, it is not uncommon for the DNA payload to become either truncated or missing from some capsids altogether. This is referred to as the empty to full capsid ratio during titering. Empty capsids still contain all of the AAV genes but none of the transgene DNA so designing a probe to anneal to the therapeutic transgene DNA is where the copy number measurement will be.

It is important when designing primers and probes for qPCR transgene amplification that species homology is taken into consideration. In most cases, the transgene of interest is a human sequence that is being used to translate a protein inside the cells for therapeutic effects. In small and large animal studies using rats, mice, dogs, or non-human primates, it must be recognized that these animals carry an endogenous sequence of these genes as well, which may be similar or dissimilar to the human sequence. If there are portions of a gene sequence that is heterologous between species, this is a good candidate for designing a probe against. A thorough probe test must be done to show the probes amplify human sequence and not the endogenous animal sequence to ensure no background copies are counted in the study qPCR readout.

7.3 qPCR inhibition

The occurrence of PCR inhibition can result in inaccurate results, which is of high concern in bioanalysis, biomarker research, and gene expression studies. Known PCR inhibitors include hemoglobin, IgG, humic substances, urea, bile salts as well as laboratory reagents including buffer or collection articles such as swabs. Inhibition of the DNA polymerase, annealing, and fluorescent detection can result in the decreased PCR efficiency. This decreased efficiency will result in the under reporting of the sample copy number in the samples and possibly the standard curve and quality controls as well. When utilizing qPCR for sample analysis, the matrix itself can result in reduction of the fluorescent signal resulting in inaccurate readouts so extraction procedures should be carefully developed to remove inhibitory factors (Sidstedt 2020; Buckwalter 2014).

Methods can be put in place to avoid inhibition and inhibition evaluation can help determine whether it has occurred. In qPCR, when a standard curve and QCs are present, the best practice is to mimic your sample composition where possible with the addition of DNA or RNA and buffer that will be used in sample extraction elution. The use of these will assure that the efficiency of the qPCR was developed with the consideration of the sample type in mind as well as potential inhibitory agents. Also, inhibition controls are often used to evaluate the present of inhibition in individual samples. Spiking a known amount of sequence specific product can be used to standardize the known amount per reaction. When a decreased CT value is shown to be detected, the sample can be reprocessed and reanalyzed. Contamination is huge concern when conducting PCR. While best practices should be employed throughout the process, there should still evaluation steps put in place to evaluate if it has occurred.

8. Data analysis: from Cq (Ct) to copy number

Real-time PCR involves continuous monitoring of the amplification process as it occurs [35]. The amplification at each cycle is monitored fluorometrically. The cycle at which the fluorescence emitted is greater than the background (in exponential phase) is taken to measure template quantity. Hence, it is called cycle quantification (Cq) or cycle threshold (Ct). Under specified conditions, the template copy number is inversely proportional to the Cq/Ct, which is plotted as log of copy number. The real-time PCR platforms are used for both detection and quantification (qPCR). RT-qPCR was originally developed from real-time PCR to amplify distinct nucleic acid sequences for the detection and relative quantification of mRNA levels in human monocyte-derived macrophages [36]. Development of single-Step RT-qPCR and efficient reverse transcriptases have made the RT-qPCR process easier and simpler [37]. In two-step RT-qPCR, the reverse transcription is separated from the real-time PCR assay. This has an advantage when several targets are to be quantified as the efficiency of reverse transcription for each template varies [38]. Though both dye (SYBR) and hydrolysis probes are in use for real-time PCR, due to specificity, multiplicity, and sensitivity, Taqman probes are preferred for quantification.

Though the digital PCR is gaining popularity and applied in several studies, still qPCR/RT-qPCR is considered as the "Gold Standard" technique due to its wide range of quantification, availability of platforms, and less expenditure. Except the conversion of RNA to cDNA by using a reverse transcriptase (RT step), the technology of qPCR and RT-qPCR remained the same. The output data from current real-time PCR platforms are Ct values. In both the gene therapy and cell therapy studies, the reporting units for absolute quantification are as follows,

a. vector genome copy number/μg of host genomic DNA. This is calculated either directly from the sample total DNA concentration or from the copy number of a reference gene [39].

b. Transcript (mRNA) copy number/μg of total RNA in the sample.

In the case of relative quantification, the units are expressed as fraction compared to the level of a selected reference gene. Relative quantification is preferred in pharmacodynamic (PD) studies for expression profiles of several biomarker gene transcripts.

Thus, to report the results in the case of absolute quantification sample DNA/RNA concentration measurement is essential. This is also essential to normalize the nucleic acid (NA) concentration in all samples to rule out the effect of this variation on Ct value.

8.1 Quantification and quality testing of total RNA in purified extracts

For total DNA/RNA quantification, the instruments, Nanodrop/UV-VIS spectrophotometer [25], Qubit, Agilent TapeStation and Promega Glomax, etc., are popularly used in different labs globally. However, specificity of Nanodrop is very less compared to the other three instruments that are fluorescent-based (Qubit & Glomax or electrophoresis based—TapeStation). In addition, Agilent TapeStation and Promega Glomax offer high-throughput technologies more suitable for the analysis of large number of samples. Nanodrop is the easiest way to test the quality of either DNA or RNA by measuring the A260/A280 nm absorbance [40]. As DNA is relatively stable, its quality is not tested once the extraction method in the matrices has been validated. However, the RNA being labile and prone to degradation easily, it is recommended to confirm the sample RNA quality before target quantitation by RT-qPCR. As the % of m-RNA in eukaryotic total RNA extracts in general will be ~5%, the quality of RNA is tested based on the integrity of 28S and 18S rRNA quantities. This is performed by one of several ways such as running on an agarose gel or TapeStation or a fragment analyzer [41].

8.2 qPCR/RT-qPCR data analysis for absolute quantification

A typical PCR-based experimental data analysis for BD of a gene therapy/cell therapy has the steps.

a. Conversion of Ct per reaction to copies per reaction—using a standard curve.

b. Copies per reaction to copies per a reference value.

c. In the case of RNA, calculation of copy number/% of DNA contamination from RT minus reaction data.

The output data from any real-time PCR instrument is Ct/Cq number. Serially diluted standards that produce a linear relationship between Ct values and initial amounts (copy number) of target DNA or RNA are used. From this, the copy number of the same target sequence in study samples based on their Ct values is drawn. In this case, the amplification efficiencies are considered same. In the case of single step RT-qPCR reactions, the standards and any Quality Controls (QC) must be RNA only [42]. GLP/regulated studies require a validated instrument and calculation sheet or a LIMS platform that can perform these calculations. Secondly, the copies/reaction are to be converted to copies per μg of DNA/RNA in the sample or mL of biofluid in clinical studies or mg of fecal material.

8.3 RT-qPCR data analysis for relative quantification of target transcript abundance

Relative quantification in gene and cell therapy involves measurement of changes in the level of sample gene expression based on an external standard or a reference sample in which the results are expressed as a target/reference ratio. A reference gene is completely different from the target gene. So, in most cases the amplification

Model name	Details
Standard curve method	Copy numbers quantitated using a standard curve and relative quantitation calculated in comparison with a selected calibrator sample
The comparative Ct method	A mathematical model that calculates changes to relative fold difference between an experimental and calibrator sample, no standard curve required
Pfaffl model	Calculates relative gene expression data while accounting for differences in primer efficiencies using the equation: Gene Expression Ratio = $(EGOI)^{\Delta Ct\,GOI}/(EHKG)^{\Delta Ct\,HKG}$ Where E = Efficiency, GOI = Gene of Interest, HKG = HouseKeeping Gene or Reference gene.
Q-gene method, Gentle et al. model, Liu and Saint method, Amplification plot method	Calculate amplification efficiencies and normalize the data.
Data Analysis for Real-time PCR (DART-PCR)	Converts raw data into R 0 values, based upon the theory that fluorescence is proportional to DNA concentration

Table 1.
Mathematical models for relative quantification of RT-qPCR data.

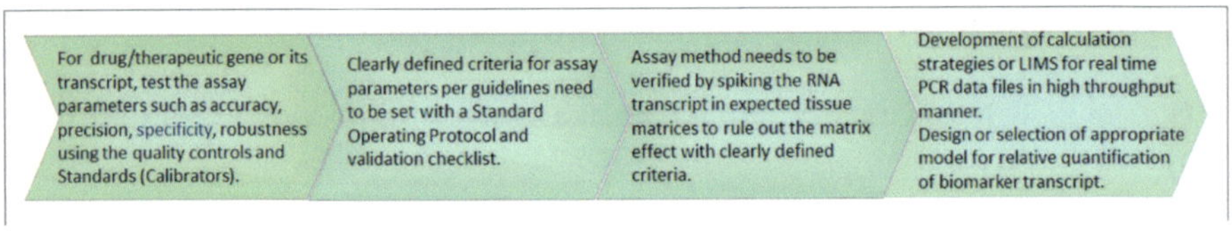

Figure 3.
PCR assay validation criteria for gene/cell therapy studies.

efficiency of the reference and target gene mRNA will be different. The simplest equation to calculate the amplification efficiency from the RT-qPCR data is $e = 10^{-1/slope}$ [43], where: e = theoretical efficiency, Slope = the slope of the standard curve, plotted with the y axis as Ct, and the x axis as log(quantity). Several mathematical models are developed to calculate the mean normalized gene expression from relative quantification assays. **Table 1** below lists some of the available models and their principles [44].

FDA has not yet required validation of qPCR/qRT-PCR assays [45]. Extensive discussion and recommendations for "best practice" for qPCR/qRT-PCR assay design during method development, and validation for gene therapy sample analysis with special focus on AAV-based clinical and non-clinical studies were made recently [46, 47]. These practices are followed for biodistribution analysis in gene/cell therapy studies for vector transgene or its mRNA quantitation by qPCR or RT-qPCR, respectively. Parameters and criteria for a typical AAV gene therapy are shown in **Figure 3**.

9. Digital PCR

Application of digital PCR (dPCR) is vastly increasing in pharma industry and basic research because of technological progress and increasingly affordable technology. The term "digital PCR" was coined by Vogelstein and Kinzler in 1999 [48]. dPCR involves dispersal of the reaction mixture into thousands of small reaction units and performing PCR with either real-time or end-point data collection in many separate reaction chambers. The partitions are performed either by droplet formation or by microfluidic nanoplate partitions. The essential steps in the data analysis of digital PCR, a) Number of partitions measured; b) Partition volume; c) Copies per partition; and d) dPCR analysis program details. The raw data of the digital PCR provide the number of positive/negative partitions (droplets in case of ddPCR) for each sample. Mean number of target sequences is calculated using the manufacturer's software applying Poisson distribution co-efficient. This also compensates for if there is more than one copy of template in some partitions. MIQE guidelines [49, 50] listed the information to be reported for the publication of dPCR data. A comparison of the digital and real-time PCR methods has been made (**Table 2**).

9.1 Real-time digital PCR systems

The popular commercial digital PCR platforms are endpoint PCR systems. This has some drawbacks primarily due to false negatives and false positives. The false-negative result happens due to weak reaction efficiency and false-positive result due to water evaporation, cross-contamination, or non-specific amplification. A real-time digital PCR system has been constructed and tested [51]. This integrated system

Item	Real time	Digital
Principle	Quantification is performed in real time as the amplification happens after every cycle. Data are collected during exponential (log) phase—to calculate the template copy number.	Partitioning of the PCR reaction into thousands of individual reaction vessels prior to amplification. Microfluidics technology is used for digitization. Mostly, data acquisition is end point. Commercial instruments for real-time digital PCR are not yet very popular.
Quantification	Both absolute and relative quantifications are possible with a set of standards and reference gene, respectively.	Absolute and relative quantifications are possible without a standard curve.
Quantification	**Cq/Ct Value** is the cycle number of PCR at which the fluorescent intensity is above the set threshold value. A standard curve is constructed with a set of standards with known copy number that gives specific Ct values.	In endpoint dPCR, the copy number is calculated based on the fraction of positive and negative droplets or partitions. A *Poisson distribution co-efficient* is used to correct the copies per partition using the equation = $-ln(1-p)$, where p = fraction of positive droplets/ partitions. The real-time dPCR reduces the false-positive/false-negative results.
Precision	Varies depending on the target and multiplicity. In general, better at higher copy number in the range.	Higher with more replicates per sample and better than real time at lower concentrations.
Upper Limit of Quantitation (ULOQ)	ULOQ is 10^8 copies for most of the targets. A maximum of 10^8 is reported.	The platforms by the year 2022 offer up to 10^5 copies of ULOQ.
Lower Limit of Quantitation (LLOQ)	Varies (3 and above copies/ rxn) depending on the target.	More consistency was reported at lower concentrations in the range compared to real-time qPCR.
Multiplexing	Higher capability	Limited
Throughput	96 or 384 well format	Only 96 well format. However, some systems can run up to 8 plates at a time.
Run Time	Can go to <90 minutes	2 hours or higher
Inhibition of PCR	Can be monitored with an internal or external positive control	The inhibitors are diluted due to partitioning/ digitization.
Cost	More economical, platforms, and trained staff are already available in most of the pharma.	Expensive initial start-up and training on the demerits/troubleshooting is warranted.

Table 2.
qPCR & dPCR—An overview.

with 3D-printed parts exhibited high accuracy in the quantification of human 18 S ribosomal RNA gene fragment cloned in a plasmid DNA compared to QuantStudio™ 3D dPCR system. In two other studies, the Gnomegen real-time digital PCR system showed superior sensitivity and accuracy compared to the end-point digital PCR system and Quantstudio 5 real-time PCR system in the detection of SARS-COV-2 RNA and in the quantification of epidermal growth factor receptor (EGFR) exon 19 deletion, T790M, and L858R point mutations as well as human epidermal growth factor receptor 2 (HER2) amplification, respectively [52, 53]. Improvement in the integrated real-time digital PCR systems and availability from other manufacturers in the future will be very helpful in the PK-Biodistribution/PD analyses for gene/cell therapy studies and for point mutation detections. Overall, sequencing did not kill PCR. CRISPR gene-edited cell therapies are also showing promising results.

Author details

Jacqueline Murphy, Kate Herr and Venkata Vepachedu*
Janssen Therapeutics, Johnson and Johnson Family of Companies, Spring House, PA, USA

*Address all correspondence to: vvepache@its.jnj.com

References

[1] Lijun Z, Cao YJ. Engineered T cell therapy for cancer in the clinic. Frontiers in Immunology. 2019;**10**:2250

[2] Available from: https://www.nobelprize.org/uploads/2018/10/press-medicine2018.pdf

[3] Haderbache R, Warda W, Hervouet E, et al. Droplet digital PCR allows vector copy number assessment and monitoring of experimental CAR T cells in murine xenograft models or approved CD19 CAR T cell-treated patients. Journal of Translational Medicine. 2021;**19**:265. DOI: 10.1186/s12967-021-02925-z

[4] Kang E, et al. Nature. 2016;**540**:270-275

[5] Wolf DP, Mitalipov PA, Mitalipov SM. Principles of and strategies for germline gene therapy. Nature Medicine. 2019;**25**:890-897. DOI: 10.1038/s41591-019-0473-85

[6] Ravichandran VA, Viswanathan P. Chapter 2. In: Biotechnology-Based Therapeutics in Translational Biotechnology. In: Yasha H, editor. London, United Kingdom: Academic Press Inc.; 2021

[7] Available from: https://www.thegene home.com/how-does-gene-therapy-work/vectors

[8] Bulcha JT, Wang Y, Ma H, et al. Viral vector platforms within the gene therapy landscape. Signal Transduction and Targeted Therapy. 2021;**6**:53. DOI: 10.1038/s41392-021-00487-6

[9] Capasso C, Garofalo M, Hirvinen M, Cerullo V. The evolution of adenoviral vectors through genetic and chemical surface modifications. Viruses. 2014;**6**(2):832-855. DOI: 10.3390/v6020832

[10] Guo X, Sun Y, Chen J, Zou X, Hou W, Tan W, et al. Restriction-assembly: A solution to construct novel adenovirus vector. Viruses. 2022;**14**(3):546. DOI: 10.3390/v14030546

[11] Sallard E, Zhang W, Aydin M, Schröer K, Ehrhardt A. The adenovirus vector platform: Novel insights into rational vector design and lessons learned from the COVID-19 vaccine. Viruses. 2023;**15**(1):204. DOI: 10.3390/v15010204

[12] Roberts D, Nanda A, Havenga M, et al. Hexon-chimaeric adenovirus serotype 5 vectors circumvent pre-existing anti-vector immunity. Nature. 2006;**441**:239-243. DOI: 10.1038/nature04721

[13] Mantwill K, Klein FG, Wang D, Hindupur SV, Ehrenfeld M, Holm PS, et al. Concepts in oncolytic adenovirus therapy. International Journal of Molecular Sciences. 2021;**22**(19):10522. DOI: 10.3390/ijms221910522

[14] Sonntag F, Schmidt K, Kleinschmidt JA. A viral assembly factor promotes AAV2 capsid formation in the nucleolus. Proceedings of the National Academy of Sciences of the United States of America. 2010;**107**:10220-10225

[15] Srivastava A. In vivo tissue-tropism of adeno-associated viral vectors. Current Opinion in Virology. 2016;**21**:75-80. DOI: 10.1016/j.coviro. 2016.08.003

[16] Wang D, Tai PWL, Gao G. Adeno-associated virus vector as a platform for gene therapy delivery. Nature Reviews. Drug Discovery. 2019;**18**:358-378. DOI: 10.1038/s41573-019-0012-9

[17] Girod A et al. Genetic capsid modifications allow efficient re-targeting of adeno-associated virus type 2. Nature Medicine. 1999;**5**:1052-1056

[18] Qian R, Xiao B, Li J, Xiao X. Directed evolution of AAV serotype 5 for increased hepatocyte transduction and retained low humoral Seroreactivity. Molecular Therapy - Methods & Clinical Development. 2020;**20**:122-132. DOI: 10.1016/j.omtm.2020.10.010

[19] Herrmann A-K, Bender C, Kienle E, Grosse S, El Andari J, Botta J, et al. A robust and all-inclusive pipeline for shuffling of adeno-associated viruses. ACS Synthetic Biology. 2019;**8**:194-206

[20] Gray SJ, Blake BL, Criswell HE, Nicolson SC, Samulski RJ, McCown TJ, et al. Directed evolution of a novel adeno-associated virus (AAV) vector that crosses the seizure-compromised blood-brain barrier (BBB). Molecular Therapy. 2010;**18**(3):570-578. DOI: 10.1038/mt.2009.292

[21] Available from: https://www.news-medical.net/life-sciences/Lentivirus-Genome.aspx#:~: text=The%20 specificity%20 of%20 the%20 Lentivirus,reverse%20 transcriptase%2C%20protease% 20and% 20integrase

[22] Vink CA et al. Eliminating HIV-1 packaging sequences from Lentiviral vector proviruses enhances safety and expedites gene transfer for gene therapy. Molecular Therapy. 2017;**25**:1790-1804

[23] Alnefaie A, Albogami S, Asiri Y, Ahmad T, Alotaibi SS, Al-Sanea MM, et al. Chimeric antigen receptor T-cells: An overview of concepts, applications, limitations, and proposed solutions. Frontiers in Bioengineering and Biotechnology. 2022;**10**:797440. DOI: 10.3389/fbioe.2022.797440

[24] Feins S, Kong W, Williams EF, Milone MC, Fraietta JA. An introduction to chimeric antigen receptor (CAR) T-cell immunotherapy for human cancer. American Journal of Hematology. 2019;**94**:S3-S9. DOI: 10.1002/ajh.25418

[25] Sterner RC, Sterner RM. CAR-T cell therapy: Current limitations and potential strategies. Blood Cancer Journal. 2021;**11**:69. DOI: 10.1038/s41408-021-00459-7

[26] Zhang FQ, Jiang JL, Zhang JT, et al. Status and future prospects of stem cell therapy in Alzheimer's disease. Neural Regeneration Research. 2020;**15**:242. DOI: 10.4103/1673-5374.265544

[27] Aly RM. Current state of stem cell-based therapies: An overview. Stem Cell Investigation. 2020;**7**:8. DOI: 10.21037/sci-2020-001

[28] Shi Y, Inoue H, Wu JC, et al. Induced pluripotent stem cell technology: A decade of progress. Nature Reviews. Drug Discovery. 2017;**16**:115-130. DOI: 10.1038/nrd.2016.245

[29] Lin HT, Okumura T, Yatsuda Y, Ito S, Nakauchi H, Otsu M. Application of droplet digital PCR for estimating vector copy number states in stem cell gene therapy. Human Gene Therapy Methods. 2016;**27**(5):197-208. DOI: 10.1089/hgtb.2016.059

[30] Bloom K, van den Berg F, Arbuthnot P. Self-amplifying RNA vaccines for infectious diseases. Gene Therapy. 2021;**28**(3-4):117-129. DOI: 10.1038/s41434-020-00204-y

[31] Hammond S, Bernstein E, Beach D, et al. An RNA-directed nuclease mediates post-transcriptional gene silencing in drosophila cells. Nature. 2000;**404**:293-296. DOI: 10.1038/35005107

[32] Yeung M, Bennasser Y, Le S, et al. siRNA, miRNA and HIV: Promises and challenges. Cell Research. 2005;**15**:935-946. DOI: 10.1038/sj.cr.7290371

[33] Lam JK, Chow MY, Zhang Y, Leung SW. siRNA versus miRNA as therapeutics for gene silencing. Molecular Therapy - Nucleic Acids. 2015;**4**(9):e252. DOI: 10.1038/mtna.2015.23

[34] Liu WL, Stevenson M, Seymour LW, Fisher KD. Quantification of siRNA using competitive qPCR. Nucleic Acids Research. 2009;**37**(1):e4. DOI: 10.1093/nar/gkn903

[35] Higuchi R, Fockler C, Dollinger G, Watson R. Kinetic PCR analysis: Real-time monitoring of DNA amplification reactions. Biotechnology (N. Y). 1993;**11**(9):1026-1030. DOI: 10.1038/nbt0993-1026

[36] Wang AM, Doyle MV, Mark DF. Quantitation of mRNA by the polymerase chain reaction. Proceedings of the National Academy of Sciences. 1989;**86**(24):9717-9721

[37] Raja S, Luketich JD, Kelly LA, et al. Increased sensitivity of one-tube, quantitative RT-PCR. BioTechniques. 2000;**29**(4):6

[38] Mannhalter C, Koizar D, Mitterbauer G. Evaluation of RNA isolation methods and reference genes for RT-PCR analyses of rare target RNA. Clinical Chemistry and Laboratory Medicine. 2000;**38**:171-177

[39] Yang TY, Doddareddy R. Considerations in the development and validation of real-time quantitative polymerase chain reaction and its application in regulated bioanalysis to characterize the cellular kinetics of CAR-T products in clinical studies. Bioanalysis. 2021;**13**(2):115-128. DOI: 10.4155/bio-2020-0221

[40] NanoDrop Spectrophotometers, Assessment of Nucleic Acid Purity. T042. Rev 1/11

[41] Comparison of RIN and RQN for the Agilent 2100 Bioanalyzer and the Fragment analyzer systems Technical Overview Agilent Technologies

[42] Giulietti A, Overbergh L, Valckx D, Decallonne B, Bouillon R, Mathieu C. An overview of real-time quantitative PCR: Applications to quantify cytokine gene expression. Methods. 2001;**25**(4):386-401

[43] Rasmussen R. Quantification on the light cycler. In: Rapid Cycle Real-Time PCR. Berlin, Heidelberg: Springer Berlin Heidelberg; 2001. pp. 21-34

[44] Čikoš Š, Bukovská A, Koppel J. Relative quantification of mRNA: Comparison of methods currently used for real-time PCR data analysis. BMC Molecular Biology. 2007;**8**:113. DOI: 10.1186/1471-2199-8-113

[45] Medicines Agency E. Guideline on the quality, non-clinical and clinical aspects of gene therapy medicinal products [Internet]. 2014. Available from: www.ema.europa.eu/contact

[46] Ma H, Bell KN, Loker RN. qPCR and qRT-PCR analysis: Regulatory points to consider when conducting biodistribution and vector shedding studies. Molecular Therapy - Methods & Clinical Development. 2021;**20**:152-168

[47] Gorovits B, Marshall JC, Smith J, Whiteley LO, Neubert H. Bioanalysis of adeno-associated virus gene therapy therapeutics: Regulatory expectations. Bioanalysis. 2019;**11**(21):2011-2024

[48] Vogelstein B, Kinzler KW. Digital PCR. Proceedings of the National Academy of Sciences of the United States of America. 1999;**96**(16):9236-9241

[49] Huggett JF, Foy CA, Benes V, et al. The digital MIQE guidelines: Minimum information for publication of quantitative digital PCR experiments. Clinical Chemistry. 2013;**59**(6):892-902

[50] Huggett JF. The digital MIQE guidelines update: Minimum information for publication of quantitative digital PCR experiments for 2020. Clinical Chemistry. 2020;**66**(8):1012-1029

[51] Zhou S, Gou T, Hu J, Wu W, Ding X, Fang W, et al. A highly integrated real-time digital PCR device for accurate DNA quantitative analysis. Biosensors & Bioelectronics. 2019;**128**:151-158. DOI: 10.1016/j.bios.2018.12.055

[52] Duong K, Ou J, Li Z, Lv Z, Dong H, Hu T, et al. Increased sensitivity using real-time dPCR for detection of SARS-CoV-2. BioTechniques. 2021;**70**(1):7-20. DOI: 10.2144/btn-2020-0133

[53] Xu J, Duong K, Yang Z, Kaji K, Ou J, Head SR, et al. Real-time digital polymerase chain reaction (PCR) as a novel technology improves limit of detection for rare allele assays. Translational Lung Cancer Research. 2021;**10**(12):4336-4352. DOI: 10.21037/tlcr-21-728

Section 4

Clinical Applications

Chapter 5

PCR Techniques and Their Clinical Applications

Mingke Wang, Jin Cai, Jinhong Chen, Jingwen Liu, Xiaoyu Geng, Xuelu Yu and Jishun Yang

Abstract

Kary B. Mullis developed a revolutionary method name polymerase chain reaction (PCR) in 1983, which can synthesize new strand of DNA complementary to the template strand of DNA and produce billions of copies of a DNA fragment only in few hours. Denaturation, annealing, and extension are the three primary steps involved in the PCR process, which generally requires thermocyclers, DNA template, a pair of primers, Taq polymerase, nucleotides, buffers, etc. With the development of PCR, from traditional PCR, quantitative PCR, to next digital PCR, PCR has become a powerful tool in life sciences and medicine. Applications of PCR techniques for infectious diseases include specific or broad-spectrum pathogen detection, assessment and surveillance of emerging infections, early detection of biological threat agents, and antimicrobial resistance analysis. Applications of PCR techniques for genetic diseases include prenatal diagnosis and screening of neonatal genetic diseases. Applications of PCR techniques for cancer research include tumor-related gene detection. This chapter aimed to discuss about the different types of PCR techniques, including traditional PCR, quantitative PCR, digital PCR, etc., and their applications for rapid detection, mutation screen or diagnosis in infectious diseases, inherited diseases, cancer, and other diseases.

Keywords: PCR, quantitative PCR, digital PCR, rapid detection, mutation screen, infectious diseases, inherited diseases, cancer

1. Introduction

The polymerase chain reaction (PCR) is one of the most widely used techniques in molecular biology, which plays an important role in traditional molecular cloning experiments and also solves the problem of obtaining sufficient amount of nucleic acid [1]. PCR technology can solve many problems in molecular biology by synthesizing primers needed for genome sequence information to amplify specific segments of DNA and has been widely used in laboratories and clinics.

In 1971, Khorana et al. first proposed the idea of nucleic acid amplification *in vitro* [2]. "tRNA genes can be synthesized by the continuous repetition of the process that DNA denaturation, hybridization with appropriate primers, and extension of primers with DNA polymerase." However, due to the undiscovered heat-stable DNA polymerase and the difficulty of primer synthesis, this idea was soon forgotten. In 1976, Jiayun Qian, a Chinese scientist, discovered stable Taq DNA polymerase, which made a fundamental contribution to the development of PCR technology [3].

In April 1983, Mullis came up with the idea for polymerase chain reaction when he was driving to his cottage in the country. In the same year, he obtained the first PCR fragment of 49 bp after 10 cycles by isotope labeling. In 1985, Mullis invented and patented PCR, with himself being the first inventor, while he worked at Cetus Corporation. In the same year, the first academic paper on PCR was published in *Science*, with Mullis as coauthor [4].

At present, PCR techniques have been widely used in life science and clinical practice to solve many problems by amplifying specific DNA fragments efficiently. Applications of PCR techniques for infectious diseases include specific or broad-spectrum pathogen detection, assessment and surveillance of emerging infections, early detection of biological threat agents, and antimicrobial resistance analysis [5, 6]. Applications of PCR techniques for genetic diseases include prenatal diagnosis and screening of neonatal genetic diseases such as Down syndrome and thalassemia [7, 8]. Applications of PCR techniques for cancer research include tumor-related genes detection, such as activating oncogenes, tumor suppressor genes, metastatic genes or metastatic suppressor genes, tumor-related viruses, and antitumor drug-resistant genes or mutation detection [9, 10].

This chapter mainly discussed about the principle, characteristics, common problems and their solutions of PCR, the different types of PCR techniques including traditional PCR, quantitative PCR (qPCR), digital PCR (dPCR), etc., and their applications for rapid detection, mutation screen or diagnosis in infectious diseases, inherited diseases, cancer, and other diseases.

2. The principle, characteristics, common problems, and their solutions of PCR

2.1 The principle of PCR

The basic principle of PCR is the synthesis of double-stranded DNA by the replication of a specific nucleic acid molecule as a template through primer extension. In addition to heat-resistant DNA polymerase, precise temperature cycle control is also a key reaction condition. PCR technology mainly includes the following three main steps: (1) Denaturation, template DNA double-strand dissociation to single strand at high temperature; (2) Annealed, specific primers and target DNA were complementary combined according to base pairing principle; and (3) Extension, using single-stranded DNA as a template, complementary DNA replicates with the extension of primers under the action of heat-resistant DNA polymerase.

The PCR reaction can be completed automatically in a special instrument, where the required PCR program can be edited. DNA denaturation temperature is usually around 94°C, and the required time is related to the structure of target DNA and PCR reaction system. Annealing temperature is affected by primer length and base composition, which is generally between 50 and 60°C. The quantity and specificity of PCR products can be improved by adjusting annealing temperature. Generally, the extension temperature is 72°C, and the reaction time depends on the length of PCR products. It is predicted that due to the longer length of PCR products, the longer reaction time will be needed (see **Figure 1A** and **B**).

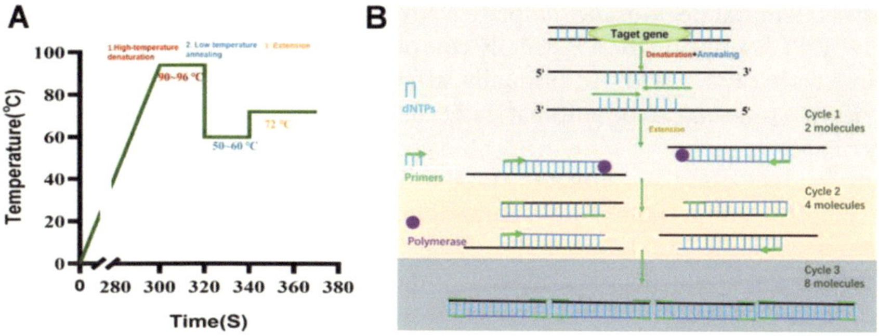

Figure 1.
Flowchart and schematic diagram of PCR technology. (A) General flowchart of PCR technology, predenaturation at 94°C for 5 min, denaturation at 90–96°C for 20 s, annealing at 50–60°C for 20 s (annealing temperature and time depends primer length and base composition), and extension at 72°C for 30 s (the reaction time depends on the length of PCR products), and (B) Schematic diagram of PCR technology, including amplification buffer, four dNTP mixtures, Taq DNA polymerase, etc., with two primers, the first 20–25 cycles of 2n index amplification, and then gradually stable after 30 cycles.

The advantage of PCR technology mainly includes its efficient amplification ability and specific replication of microtarget nucleic acid sequences. This simple technical design enables PCR technology to be widely used and play an important role in life science and related scientific fields [11].

2.2 The characteristics of PCR

The PCR reaction system includes template, primer, enzyme, dNTPs, and buffer. The template is the target DNA, which must meet the requirements of proper concentration and high purity. Too low or too high concentration of the template may lead to amplification failure, and low purity may affect the amplification efficiency. There is a pair of primers in PCR reaction, including upstream primer and downstream primer. The upstream primer is complementary to a short DNA sequence at the 5′ end of the single strand of DNA, and the downstream primer is complementary to a short DNA sequence at the 3′ end of this segment. Factors should be considered when designing primers such as its length, base composition, cross-complementary sequence within and between primers, specificity of primers, and continuous pairing at the template of 3′ ends. There are two kinds of heat-resistant DNA polymerase, that is, Taq and Pfu, from two different thermophiles [12]. In Taq enzyme, amplification efficiency is high but easy to mismatch, while in Pfu enzyme, amplification efficiency is lower but has error correction function. Deoxyribonucleoside triphosphate (dNTP) is a premixed solution containing deoxyadenosine triphosphate (dATP), dCTP, deoxyguanosine triphosphate (dGTP), and deoxythymidine triphosphate (dTTP) [13]. The composition of buffer solution is more complex. Besides water, the buffer system usually includes N-2-hydroxyethylpiperazine-N-ethane-sulphonicacid (HEPES) or 4-Morpholinepropanesulfonic acid (MOPS) buffer system [14] (univalent cation, generally using potassium ion; divalent cation, mainly magnesium ion; and auxiliary components, commonly using dimethylsulfoxide (DMSO) and glycerin), which is mainly used to maintain enzyme activity and help DNA unwinding structure.

PCR amplification follows the principle of base complementary pairing, which has high specificity in theory. PCR product synthesis increases exponentially and amplifies the initial peak-grade template DNA to the microgram level with high sensitivity. PCR amplification is usually completed within 2–4 hours, which has the superiorities in saving time. Additionally, after the PCR reaction program is set, the amplification process can be automatically completed on the PCR instrument, which is easy to operate.

Four factors mainly determine the PCR reaction time: heating and cooling time of the template, temperature equilibrium time between the template and the PCR tube, extension time, and number of cycles [15]. Based on the principle of PCR, three temperature points of denaturation, annealing, and extension were set in three steps. The double-stranded DNA was denatured at 94°C and then rapidly cooled to 50–60°C; the primer was annealed and bound to the target sequence and then heated to 72°C. Under the action of Taq polymerase, the primer chain was extended along the template [16]. Among the four main factors, the extension time and the number of cycles play a more decisive role in the reaction time of PCR.

The number of PCR cycles is usually about 30–40. Although the number of products can be increased, the amplification of nonspecific products will also be increased correspondingly as the number of cycles increased. Additionally, the polymerase activities in the PCR reaction system are limited due to the temperature variation from high to low in the recirculation, and the amount of primers and raw materials in the PCR system is not infinite [17]. With the consumption of dNTP, primers and other raw materials in the PCR reaction system and the decrease of enzyme amplification ability, mismatch, and nonspecific amplification will also be increased as the number of cycles increased. Therefore, if the PCR products are sufficient for subsequent analysis, it is generally less likely to extend the number of PCR cycles excessively.

2.3 The common problems and their solutions of PCR

There are many possible reasons for false negative in PCR [18], and the corresponding treatment methods are different. First, the template was lost or degraded too much or contained too much protein and many impurities, which may be solved by the extraction quality of the template. Second, the PCR reaction system was prepared incorrectly, which may be solved by the reprepared experiment. Third, the DNA polymerase was inactivated or the concentration was too low, which may be solved by the replaced DNA polymerase and operated on the ice bath. Fourth, the PCR reaction procedure was not appropriately set, which may be solved by the adjusted PCR reaction conditions, such as reduction of denaturation time or annealing temperature. Fifth, the primer design was unreasonable or the synthesis quality was not high. In this case, primer design software should be used to further simulate and analyze its amplification quality or redesign and synthesize the primers if necessary.

There are relatively possible reasons for few false positive in PCR. First, primers have low specificity and certain homology with nontarget amplified sequences [19]. In this situation, PCR reaction conditions can be adjusted, such as increasing annealing temperature and redesigning the primers if necessary. Second, target DNA or amplified products were cross-contaminated [20]. In those circumstances, consumables such as PCR reaction tubes, centrifuge tubes, and pipette tips should be used once as far as possible. Additionally, before operation, the air in the laboratory, reaction reagent, or sampling gun could be exposed under ultraviolet light. Moreover, operation should be careful and gentle to prevent target DNA template from being sucked into the sampling gun or spattered out of the centrifuge tube.

3. The history of PCR development

3.1 The first generation: ordinary PCR

Usually what we call the first generation of ordinary PCR refers to the original PCR technology, which used the application of the ordinary PCR machine to amplify the target DNA and then analyzed PCR products by sequencing or agarose gel electrophoresis, and was generally used for qualitative analysis [21]. It was also used for semiquantitative analysis of gene transcription expression levels, before the second generation developed [22].

The first generation can effectively amplify trace amounts of DNA, especially in paleontological remains in fossils and samples coming from hair, skin debris, blood, or body fluids at crime scenes [23, 24]. Only a small number of DNA sequences were needed to be extracted, while enough copies can be obtained after amplification using PCR. Further sequencing analysis and homology comparison can smoothly go on.

In addition, the PCR products were analyzed by agarose gel electrophoresis for the amount, size, brightness of stripes, etc. According to the information from electrophoresis band, the target DNA fragment was prepared, the gene to be tested was qualitatively analyzed, and the gene expression was semiquantitatively analyzed [25]. However, a major drawback existed in the first generation of PCR technology, which was generally limited to qualitative analysis and unable to achieve accurate quantitative analysis. Besides, the nucleic acid dyes used at the beginning of agarose gel electrophoresis, such as ethidium bromide, are chemical carcinogens and have great harm to human body and environment [26]. And the detection process of sequencing or electrophoresis analysis takes quite a long time [27].

3.2 The second generation: quantitative PCR

The second generation of quantitative PCR (qPCR) refers to real-time PCR technology. qPCR adds fluorescent dyes or DNA probes into the reaction system to indicate the reaction progression and collects fluorescence signals to monitor PCR product accumulation. By analyzing the cycle threshold value (Ct value) of the amplification curve, the amount of PCR product is quantified in combination with the standard curve [28]. This technology can bring the results to repeatability and comparability, after being performed in accordance with standard operating procedures [29].

This technology is widely used in the medical field, due to the advantages of quantitative analysis, lower risk of cross-contamination, and easier operation. It has played an important role in the clinical application for infectious disease pathogen nucleic acid detection, anti-infection, antitumor and other drug resistance gene detection, genetic disease diagnosis, gene therapy, and so on [30–34].

qPCR overcomes the disadvantages of the first generation that cannot achieve accurate quantitative analysis, but it is still a relatively quantitative analysis method, which relies heavily on the level of operation technology and standard curve quality. For example, low-copy target DNA is difficult to detect, and background values skew test results and are greatly affected by PCR inhibitors [35]. Also, there are certain differences between the reagents and equipment produced by different manufacturers [36].

3.3 The third generation: Digital PCR

As next generation of PCR technology, digital PCR (dPCR) dispenses nucleic acid samples into a large number of independent, parallel microreaction units (nanoliters), then dilutes the template in each reaction unit to the level of single molecules, and carries out amplification, detection, and distribution statistics on this basis, so as to achieve absolute counting of target molecules. According to its reaction unit carrier, dPCR is further divided into droplet digital PCR based on droplet microfluidics and chip microarray PCR based on chip microfluidics [37].

This technology reduces the concentration of nonspecific background sequences greatly, reduces amplification bias largely, and has low background values and strong specificity. Low-copy target DNA can be detected with high sensitivity due to small effects of template concentration. The quantitative analysis of target DNA does not require drawing a standard curve, which can achieve absolute quantification with good accuracy [38, 39]. Furthermore, this technology has obvious advantages in the reproducibility of reagents from different manufacturers and operator test results because of fewer interference factors.

dPCR is especially suitable for nucleic acid trace analysis with very high sensitivity requirements and accurate quantification of nucleic acids in complex samples such as tissues, body fluids, and excreta. For example, the accurate analysis of small changes in viral load is critical for determining the course of the disease and assessing efficacy [40]. Also, dPCR can accurately quantify mutation frequencies as low as 0.001–0.0001% due to its ultrahigh sensitivity, which can be used in the detection of rare mutation sites of genes related to anti-tumor-targeted drugs. dPCR can enrich target genes in the sample to be sequenced, validate, and accurately quantify sequencing results [41]. It has been reported of successfully applications for researches in single-cell or low-abundance gene expression analysis, gene expression fold change less than two times, and copy number variation with gene copy number greater than 5 [42–44].

However, dPCR has strict requirements for the amount of template, which in each reaction unit need to be diluted to the single-molecule level by limiting allocation, too much will lead to inability to quantify, and too little will lead to low signal. In addition, PCR amplification primers require a high specificity, and nonspecific amplification should be avoided as much as possible.

In addition to the PCR techniques described above, various methods have been developed to improve PCR. Examples include multiplex PCR, broad-range PCR, strip PCR, and direct strip PCR methods, which are used to detect 24 pathogens that cause ocular infections [45]. Oligoribonucleotide interference PCR (ORNi-PCR) can amplify the desired transcript variant in cDNA by inhibiting the amplification of unwanted transcripts and can simultaneously detect two single-nucleotide mutations in the same allele of epidermal growth factor receptor (EGFR) gene in lung cancer cells with high sensitivity, which is expected to be used for the early detection of cancers caused by single-nucleotide mutations [46]. Furthermore, there are other types of PCR, such as methylation-specific PCR, type-specific PCR, etc. [47, 48].

4. Clinical applications of PCR

4.1 Applications of PCR for infectious diseases

At present, PCR is widely used in clinical practice, especially in the detection, screening, or diagnosis of infectious diseases. Its applications include specific or broad-spectrum pathogen detection, assessment and monitoring of emerging infections, detection of biological threat agents early, and analysis of antimicrobial resistance [6]. The main methods for routine diagnosis of pathogens include bacterial culture, virus culture, cell culture, serological diagnosis, and enzyme-linked immunosorbent assay. However, these detection methods have different disadvantages, such as tedious process of traditional cell culture and pathogen isolation and identification, time-consuming operation, high price, and low sensitivity [49, 50]. And serological diagnosis is prone to false positive results due to cross-linking reaction [51]. Compared with the above detection methods, the nucleic acid molecular detection method based on the nucleic acid amplification principle can significantly improve the sensitivity and specificity of detection. PCR technology breaks through the "time window" limit of immunological detection and can judge whether the disease is in recessive or subclinical state [52, 53]. Therefore, considering that one of the key points of prevention and treatment of infectious diseases is early and rapid diagnosis, PCR technology can overcome the lag of immunological detection and the complexity of gene sequencing. Since the epidemic of novel coronavirus infection (coronavirus disease 2019—COVID-19) for more than 3 years, PCR nucleic acid detection has always been the primary method for the diagnosis of infection [5]. With the development of PCR technology, multiplex PCR, multiplex quantitative PCR, microfluidic PCR, and other detection platforms have been developed to detect multiple pathogens at the same time [54–56].

In addition, PCR is used to identify pathogenic microorganisms by typing. For example, when testing for human papillomavirus (HPV), PCR can not only identify the infection, but also the subtype to which it belongs [57]. The pathogenicity of different HPV subtypes varies significantly. Assessing cancer risk based on subtype distribution and adopting individualized and precise prevention and control strategies can prevent cervical cancer more effectively [58].

PCR techniques also allow for the quantification of pathogenic microbial loads. For example, viral load is an important indicator for assessing efficacy of antiviral drug therapy. PCR is used to quantify changes in the load levels of hepatitis B virus (HBV) or human immunodeficiency virus (HIV) in blood to determine the effectiveness of treatment for HBV or HIV [59].

There are also more established applications of PCR techniques in antimicrobial resistance genetic testing. For example, *Klebsiella pneumoniae* is one of the important causative agents of hospital infections. With the widespread use of broad-spectrum antimicrobial drugs, such as β-lactams and aminoglycosides, bacteria are prone to produce extended-spectrum β-lactamases (ESBLs) and cephalosporinase (AmpC enzyme) as well as aminoglycoside-modifying enzymes (AMEs), resulting in drug resistance [60–62]. The drug resistance genes of clinical isolates of *K. pneumoniae* were quickly detected by PCR. Compared with the drug sensitivity test based on bacterial culture, it has unique advantages in guiding the rational use of antimicrobials [63, 64].

The third-generation digital PCR (dPCR) can detect trace drug resistance mutations of pathogenic microorganisms and achieve "accurate attack" in the selection of effective anti-infective drugs [65]. In view of the H275Y point mutation of neuraminidase H275Y in 2009 H1N1 influenza virus resistant to oseltamivir, Taylor and co-workers compared the ability of the second-generation fluorescent quantitative PCR (qPCR) and dPCR to detect H275Y under the background of high-abundance wild-type H1N1. The results showed that dPCR significantly improved the sensitivity of mutant virus detection (>30 times) [66]. Additionally, Mukaide and co-workers established a dPCR method to detect mutations in the core amino acid 70th site of hepatitis C virus (HCV) with 200-fold higher detection rate than qPCR (**Table 1**) [67].

4.2 Applications of PCR for inherited diseases

Traditional prenatal gene diagnosis mainly depends on the probe-based Southern blotting and restriction fragment length polymorphism (RFLP) to detect gene mutation including gene deletion and mismatch. However, due to its complex operation, high technical requirements, isotope labeling, and long time, its application is limited. After the rapid development of PCR in the late 1980s, it has become a common technique for the prenatal genetic diagnosis of genetic diseases. For example, PCR technology is applied to detect amniotic fluid, villi, and other samples for the rapid prenatal diagnosis of Down syndrome, thalassemia, and other genetic diseases.

Down syndrome, also known as trisomy 21 syndrome, can cause mental retardation and severe birth defects in newborns [69]. Traditional fetal aneuploidy screening is usually used in the first and second trimesters by serum biochemical testing (SBT). High-risk pregnancies identified by SBT are referred for invasive surgery, which is not only time consuming but also a risk of miscarriage. In addition, SBT can only indirectly estimate the risk or probability of pregnancy carrying aneuploid fetus based on alternative serological markers, which is not only a relatively low sensitivity, but also a high false positive rate [70]. dPCR-NIPT assay has better sensitivity and specificity than SBT, and the cost is much lower than NGS-NIPT. The simultaneous detection of fetal T21 in maternal plasma samples in a single reaction would be a better solution as the primary screening technique for fetal euploidy screening [7].

Thalassemia is an inherited blood disorder, with alpha-thalassemia and beta-thalassemia being the most common [71]. The prenatal diagnosis of monogenic diseases, such as beta-thalassemia, currently relies on invasive surgery. Since the discovery of cell-free fetal DNA (cffDNA) in maternal plasma, noninvasive prenatal diagnosis (NIPD) has made remarkable progress [72]. Guizhou Province took the lead in using next-generation sequencing (NGS) combined with the traditional screening technique for thalassemia—cross break point PCR (also named as Gap-PCR) is used

Year	PCR type	Clinical application	Reference
2010	TS-PCR	Detection of HPV subtypes (types 6, 11, 16, 18, 31, 33, and 35)	[58]
1996	PCR	Detection of the effect of lamivudine on HBV replication in HIV infected patients	[59]
2008	RAPD-PCR	Detection of *Klebsiella pneumoniae* (Molecular typing)	[63]
2016	PCR	Detection of *K. pneumoniae* (extended-spectrum β-lactamase (ESBL) producers)	[64]
2015	ddPCR	H1N1pandemic (detection of oseltamivir-resistant subpopulations)	[66]
2014	ddPCR	Detection of HCV (mutation at core amino acid 70)	[67]
2020	RT-qPCR	Detection of COVID-19 virus	[68]

Note: TS-PCR: type-specific polymerase chain reaction; PCR: polymerase chain reaction, RAPD-PCR: random amplification of polymorphic DNA-polymerase chain reaction; ddPCR: droplet digital polymerase chain reaction; and RT-qPCR: real-time-quantitative polymerase chain reaction.

Table 1.
Application of PCR for infectious diseases.

to screen newborn thalassemia genes, suggesting that NGS-Gap-PCR can effectively identify new mutations and reduce the misdiagnosis rate [8]. Microdrop digital PCR (ddPCR) is an efficient, simple, and cost-effective PCR-based technique that enables absolute quantification of target DNA molecules. D 'Aversa et al. demonstrated for the first time that a ddPCR-based NIPT method can be used to rapidly and sensitively identify materially inherited beta IVSI-110 and beta +039 thalassemia mutations [73]. Constantinou et al. optimized ddPCR assay to improve the sensitivity and specificity of monitoring and used it to determine fetal genotypes in couples noninvasively at risk of IVSI-110G > A β-thalassemia. It was suitable for all pregnancies of mothers of carriers of IVSI-110G > A, showing 97.06% accuracy [74]. In addition, PCR can also be used for NIPD of single gene diseases such as deafness and cystic fibrosis, and the results show that NIPD is accurate and effective and has the advantages of simple and rapid [75–77].

Screening for neonatal genetic diseases enables children to be diagnosed and treated within 10 days after birth, to prevent death, disability, or reduce the degree of disability due to genetic diseases. At present, qPCR and gene sequencing are the main technologies for screening neonatal genetic diseases. Genetic screening disorders range from a single disorder, such as hereditary deafness, to multiple disorders [78]. Hearing screening combined with genetic screening is helpful for the early diagnosis and intervention of clinical hearing impairment. Genetic screening for deafness can improve the detection rate of hereditary deafness. The current screening method is quantitative PCR for several genes and mutation sites, which has achieved remarkable results [79]. Garbuz et al. developed a genetic diagnostic method for Wilson's disease (WD) based on ARMS-PCR, DreamTaq green PCR premix, and modified primers to detect 14 mutations in Europeans. This diagnostic method is rapid and easy to use and has good accuracy and repeatability of mutation detection (**Table 2**) [80].

4.3 Applications of PCR for cancer and other diseases

Tumor-related genes include oncogenes, tumor suppressor genes, and genes that promote or inhibit metastasis. At present, in the field of clinical diagnosis and treatment of cancer, PCR technology is mainly applied to the detection of tumor-related genes, such as chromosome translocation [81], activated oncogene [82], tumor suppressor gene [83], metastasis gene or metastasis suppressor gene [84], tumor-related virus and antitumor drug resistance gene or mutation site [84–87], etc. By detecting gene mutation and activation expression, it can provide evidence-based medicine for analyzing tumor etiology, diagnosis and classification, metastasis, surgical treatment indication, selecting the best chemotherapy scheme, evaluating the curative effect, and judging the prognosis.

About 95% of chronic myelogenous leukemia (CML) have Philadelphia chromosome positive, which is the result of translocation of the long arm of chromosome 9 and 22. This translocation rearranges the proto-oncogene abl on the long arm of chromosome 9 (9q34) and the bcr gene on chromosome 22 (22q11) to form a fusion gene, known as Bcr-Abl fusion gene [88]. Overexpression and activation of this gene is considered to be one of the leading causes of CML, which can precede clinical symptoms, so it can be used as a basis for early diagnosis. In addition, it is known that patients with CML with negative chromosome in Philadelphia have a poor response to treatment and poor prognosis. At present, the detection of Bcr-Abl fusion gene by PCR has been deeply applied in the clinical diagnosis and treatment of CML [9].

Year	PCR type	Clinical application	Reference
2022	dPCR	prenatal diagnosis of Down syndrome	[7]
2021	PCR	prenatal diagnosis of thalassemia	[8]
2022	ddPCR	prenatal diagnosis of β-thalassemia	[73]
2022	ddPCR	prenatal diagnosis of β-thalassemia	[74]
2018	dPCR	prenatal diagnosis of deafness	[75]
2019	Fluorescent multiplex PCR	prenatal diagnosis of cystic fibrosis	[76]
2021	ddPCR	prenatal diagnosis of β-thalassemia	[77]
2019	qPCR	genetic screening for deafness	[79]
2022	PCR	genetic testing for Wilson's disease	[80]

Note: PCR: polymerase chain reaction; dPCR: digital polymerase chain reaction; ddPCR: droplet digital polymerase chain reaction; and qPCR: quantitative polymerase chain reaction.

Table 2.
Application of PCR for genetic diseases.

The study found that miR-122 is closely related to the classification of hepatocellular carcinoma [89]. The inhibition of miR-122 expression promotes the occurrence of hepatocellular carcinoma, which is characterized by enhanced invasion, easy recurrence, and reduced survival time [90]. Similarly, miR-1258 has also been reported to act as a tumor suppressor in liver cancer. miR-1258 inhibits the progress of breast cancer (BC) by increasing cell apoptosis and reducing cell viability and cell cycle transformation by inactivating key protein (KDM7A). Unlike miR-122, miR-1258 is also involved in gastric cancer (GC), colorectal cancer (CRC), oral squamous cell carcinoma (OSCC), and other diseases [91–93]. The detection of miR-122/1258 expression level by PCR can provide important reference for the classification and prognosis of hepatocellular carcinoma [10].

Ovarian cancer is a highly malignant tumor, which is insidious and progresses rapidly. Only about one-quarter of ovarian cancer is found in the early stage, and most patients have spread or metastasized at the time of treatment, so the early diagnosis of ovarian cancer is very important. In recent years, it has been found that human epididymis protein 4 (HE4) is an important biomarker for judging benign and malignant ovarian cancer [94]. Using PCR technology, the increased expression of HE4 before 4–5 months of the recurrence in advanced ovarian cancer after surgical treatment was found, suggesting that HE4 can be used to detect the recurrence of ovarian cancer.

Colorectal cancer (CRC) has also become increasingly common in recent years. Only in 2018, there were more than 1.8 million new cases of colorectal cancer and more than 880,000 deaths. Among all cancer cases in the world, CRC ranked third in the world in terms of incidence rate and the second in terms of mortality [95]. It is expected that by 2030, the global CRC burden will increase by 60%, with more than 2.2 million new cases and more than 1.1 million deaths [96]. The application of PCR in CRC is based on the fact that the microbiota spectrum associated with CRC is different from that of healthy subjects and is related to different mucosal gene expression profiles. The quantitative analysis of the host gene expression involved in CRC progression and immune response is performed by real-time quantitative PCR [97].

During tumor treatment, the dynamic detection of tumor-related gene expression in blood has obvious advantages for effect evaluation, individualized treatment, and

Year	PCR type	Clinical application	Reference
2023	PCR	Prostatic cancer, oophoroma (HE4)	[98]
2020	qRT-PCR	Lung cancer (TERT, HK-2, EGFR, KRAS)	[99]
2023	qRT-PCR	Identification of differentially expressed circRNA in GC of gastric cancer	[100]
2021	Digital droplet PCR (ddPCR)	Colorectal cancer (EVL, NTRK3)	[101]
2008	Methylation specific PCR	Prostate cancer (ASC, CDH13)	[102]
2008	Methylation specific PCR	Breast cancer (PITX2)	[103]
2013	Methylation specific PCR	Ovarian cancer (APC, RASSF1A, CDH1, RUNX3, TFPI2, SFRP5, and OPCML)	[104]
2013	Methylation specific PCR	Lung cancer (p16INK4A, CDH13, APC, DAPK)	[105]

Note: PCR: polymerase chain reaction; qRT-PCR: quantitative real-time polymerase chain reaction; and ddPCR: droplet digital polymerase chain reaction.

Table 3.
Application of PCR technology for cancer research.

metastasis. Especially for patients with nonsolid tumors and postoperative chemotherapy, there is no need to measure solid lesions. This method is particularly important for evaluating the sensitivity of chemotherapy (**Table 3**).

Additionally, PCR is also widely used in the diagnosis and treatment of other diseases, such as in parasitic diseases [106], metabolic diseases [107–109], autoimmune diseases (rheumatoid arthritis and diabetes) [110], pediatric pulmonary diseases [111], spinal muscular dystrophy diseases [112], and so on. In a word, PCR has been used in almost the whole process of disease prevention, diagnosis, treatment, and prognosis.

5. Conclusion

In conclusion, PCR is a powerful technique for obtaining nucleic acid quickly and accurately, which can synthesize new DNA strands that complement the DNA template strands and produce billions of DNA fragments. This chapter reviews the principle, characteristics, common problems, and solutions of PCR, different types of PCR techniques, including traditional PCR, quantitative PCR, digital PCR, and their rapid detection, mutation screening or diagnostic applications in infectious diseases, genetic diseases, cancer, and other diseases. Currently, PCR has become a powerful tool in the field of life science and medicine. Based on PCR techniques, detection systems with higher sensitivity and specificity are constantly developed. In the future, with the continuous upgrading of PCR, it will play an important role in molecular diagnostic technology and have a wider application prospect in clinical and more fields. Further studies on PCR techniques and their clinical application not only provide theoretical basis for the analysis of disease etiology and the prevention of disease development, but also provide new ideas for the indication of surgical treatment and the selection of the best disease treatment plan through the detection of genes mutation or differential gene expression.

Acknowledgements

This work was supported by Scientific Research Project of Shanghai Municipal Health Commission (No. 202140061), and Project Excellence of Scientific and Technological Innovation of Naval Medical Center (No. 21TPZY3701 and No. 21TPZYL01).

Conflict of interest

The authors declare no conflict of interest.

Author details

Mingke Wang†, Jin Cai†, Jinhong Chen, Jingwen Liu, Xiaoyu Geng, Xuelu Yu and Jishun Yang*
Naval Medical Center of PLA, Naval Medical University, Shanghai, China

*Address all correspondence to: jasunyang@foxmail.com

† Wang MK and Cai J contributed equally to this work.

References

[1] Mullis KB, Faloona FA. Specific synthesis of DNA in vitro via a polymerase-catalyzed chain reaction. Methods in Enzymology. 1987;**155**:335-350

[2] Khorana HG. Total synthesis of the gene for an alanine transfer ribonucleic acid from yeast. Pure and Applied Chemistry. 1971;**25**:91-118

[3] Chien A, Edgar DB, Trela JM. Deoxyribonucleic acid polymerase from the extreme thermophile Thermus aquaticus. Journal of Bacteriology. 1976;**127**:1550-1557

[4] Saiki RK, Scharf S, Faloona F, Mullis KB, Horn GT, Erlich HA, et al. Enzymatic amplification of beta-globin genomic sequences and restriction site analysis for diagnosis of sickle cell anemia. Science. 1985;**230**:1350-1354

[5] Huergo MAC, Thanh NTK. Current advances in the detection of COVID-19 and evaluation of the humoral response. Analyst. 2021;**146**:382-402

[6] Yang S, Rothman RE. PCR-based diagnostics for infectious diseases: Uses, limitations, and future applications in acute-care settings. The Lancet Infectious Diseases. 2004;**4**:337-348

[7] Dai P, Yang YF, Zhao GY, Gu ZQ, Ren HN, Hu S, et al. A dPCR-NIPT assay for detections of trisomies 21, 18 and 13 in a single-tube reaction-could it replace serum biochemical tests as a primary maternal plasma screening tool? Journal of Translational Medicine. 2022;**20**:269

[8] Tan M, Bai Y, Zhang X, Sun J, Huang C, Tian R, et al. Early genetic screening uncovered a high prevalence of thalassemia among 18 309 neonates in Guizhou, China. Clinical Genetics. 2021;**99**:704-712

[9] Atallah E, Schiffer CA, Radich JP, Weinfurt KP, Zhang MJ, Pinilla-Ibarz J, et al. Assessment of outcomes after stopping tyrosine kinase inhibitors among patients with chronic myeloid Leukemia: A nonrandomized clinical trial. JAMA Oncology. 2021;**7**:42-50

[10] Huang T, Jiang SW, Qin L, Senkowski C, Lyle C, Terry K, et al. Expression and diagnostic value of HE4 in pancreatic adenocarcinoma. International Journal of Molecular Sciences. 2015;**16**:2956-2970

[11] Analytical Methods Committee Amctb N. PCR - the polymerase chain reaction. Analytical Methods. 2013;**6**:333-336

[12] Cline J, Braman JC, Hogrefe HH. PCR fidelity of pfu DNA polymerase and other thermostable DNA polymerases. Nucleic Acids Research. 1996;**24**:3546-3551

[13] Liu C, Feng JN, Li WW, Zhu WW, Xue YX, Wang D, et al. Maintenance of dNTP pool homeostasis and genomic stability. Yi Chuan. 2022;**44**:96-106

[14] Ahmad A, Ghasemi J. New buffers to improve the quantitative real-time polymerase chain reaction. Bioscience, Biotechnology, and Biochemistry. 2007;**71**:1970-1978

[15] Green MR, Sambrook J. Polymerase chain reaction. Cold Spring Harbor Protocol. 2019;**2019**:436-456

[16] Lorenz TC. Polymerase chain reaction: Basic protocol plus troubleshooting and optimization

strategies. Journal of Visualized Experiments. 2012:e3998

[17] Freeman WM, Walker SJ, Vrana KE. Quantitative RT-PCR: Pitfalls and potential. BioTechniques. 1999;**26**(112-22):124-125

[18] Tvorogova MG, Gushchin AE. PCR studies in laboratory diagnosis: Problems of quality assurance. Klinicheskaia Laboratornaia Diagnostika. 2006;**2006**:38-41

[19] Lo YT, Shaw PC. DNA barcoding in concentrated Chinese medicine granules using adaptor ligation-mediated polymerase chain reaction. Journal of Pharmaceutical and Biomedical Analysis. 2018;**149**:512-516

[20] Sasan MS. False positive PCR for herpes simplex virus (HSV) in childhood CNS infections. Iranian Red Crescent Medical Journal. 2012;**14**:197

[21] Yamaguchi K, Ushijima H, Tsuchie H, Matsumoto T, Matsuda J, Kitamura T. Novel PCR methods for detection of HIV proviruses using digoxigenin labelled probes--Southern blot and dot blot. Kansenshōgaku Zasshi. 1991;**65**:267-272

[22] Golay J, Passerini F, Introna M. A simple and rapid method to analyze specific mRNAs from few cells in a semi-quantitative way using the polymerase chain reaction. PCR Methods and Applications. 1991;**1**:144-145

[23] Loreille O, Orlando L, Patou-Mathis M, Philippe M, Taberlet P, Hanni C. Ancient DNA analysis reveals divergence of the cave bear, Ursus spelaeus, and brown bear, Ursus arctos, lineages. Current Biology. 2001;**11**:200-203

[24] Alaeddini R. Forensic implications of PCR inhibition--A review. Forensic Science International. Genetics. 2012;**6**:297-305

[25] Poddar SK. Influenza virus types and subtypes detection by single step single tube multiplex reverse transcription-polymerase chain reaction (RT-PCR) and agarose gel electrophoresis. Journal of Virological Methods. 2002;**99**:63-70

[26] Sonmezoglu OA, Ozkay K. A new organic dye-based staining for the detection of plant DNA in agarose gels. Nucleosides, Nucleotides & Nucleic Acids. 2015;**34**:515-522

[27] Long K, Cai L, He L. DNA sequencing data analysis. Methods in Molecular Biology. 2018;**1754**:1-13

[28] Postollec F, Falentin H, Pavan S, Combrisson J, Sohier D. Recent advances in quantitative PCR (qPCR) applications in food microbiology. Food Microbiology. 2011;**28**:848-861

[29] Redshaw N, Wilkes T, Whale A, Cowen S, Huggett J, Foy CA. A comparison of miRNA isolation and RT-qPCR technologies and their effects on quantification accuracy and repeatability. BioTechniques. 2013;**54**:155-164

[30] Welti M, Jaton K, Altwegg M, Sahli R, Wenger A, Bille J. Development of a multiplex real-time quantitative PCR assay to detect chlamydia pneumoniae, legionella pneumophila and mycoplasma pneumoniae in respiratory tract secretions. Diagnostic Microbiology and Infectious Disease. 2003;**45**:85-95

[31] Xu L, Xia H, Ni D, Hu Y, Liu J, Qin Y, et al. High-dose dexamethasone manipulates the tumor microenvironment and internal metabolic pathways in anti-tumor progression. International Journal of Molecular Sciences. 2020;**21**:1846

[32] Ferreira ID, Rosario VE, Cravo PV. Real-time quantitative PCR with SYBR Green I detection for estimating copy numbers of nine drug resistance candidate genes in plasmodium falciparum. Malaria Journal. 2006;**5**:1

[33] Chen HF, Chang SP, Wu SH, Lin WH, Lee YC, Ni YH, et al. Validating a rapid, real-time, PCR-based direct mutation detection assay for preimplantation genetic diagnosis. Gene. 2014;**548**:299-305

[34] Wang F, Puddy AC, Mathis BC, Montalvo AG, Louis AA, McMackin JL, et al. Using QPCR to assign infectious potencies to adenovirus based vaccines and vectors for gene therapy: Toward a universal method for the facile quantitation of virus and vector potency. Vaccine. 2005;**23**:4500-4508

[35] Demeke T, Jenkins GR. Influence of DNA extraction methods, PCR inhibitors and quantification methods on real-time PCR assay of biotechnology-derived traits. Analytical and Bioanalytical Chemistry. 2010;**396**:1977-1990

[36] Svec D, Tichopad A, Novosadova V, Pfaffl MW, Kubista M. How good is a PCR efficiency estimate: Recommendations for precise and robust qPCR efficiency assessments. Biomolecular Detection Quantify. 2015;**3**:9-16

[37] Cao L, Cui X, Hu J, Li Z, Choi JR, Yang Q, et al. Advances in digital polymerase chain reaction (dPCR) and its emerging biomedical applications. Biosensors & Bioelectronics. 2017;**90**:459-474

[38] Dobnik D, Spilsberg B, Bogozalec Kosir A, Holst-Jensen A, Zel J. Multiplex quantification of 12 European Union authorized genetically modified maize lines with droplet digital polymerase chain reaction. Analytical Chemistry. 2015;**87**:8218-8226

[39] Pinheiro LB, Coleman VA, Hindson CM, Herrmann J, Hindson BJ, Bhat S, et al. Evaluation of a droplet digital polymerase chain reaction format for DNA copy number quantification. Analytical Chemistry. 2012;**84**:1003-1011

[40] Hijano DR, Brazelton de Cardenas J, Maron G, Garner CD, Ferrolino JA, Dallas RH, et al. Clinical correlation of influenza and respiratory syncytial virus load measured by digital PCR. PLoS One. 2019;**14**:e0220908

[41] Jiang XW, Liu W, Zhu XY, Xu XX. Evaluation of EGFR mutations in NSCLC with highly sensitive droplet digital PCR assays. Molecular Medicine Reports. 2019;**20**:593-603

[42] O'Hara R, Tedone E, Ludlow A, Huang E, Arosio B, Mari D, et al. Quantitative mitochondrial DNA copy number determination using droplet digital PCR with single-cell resolution. Genome Research. 2019;**29**:1878-1888

[43] Millier MJ, Stamp LK, Hessian PA. Digital-PCR for gene expression: Impact from inherent tissue RNA degradation. Scientific Reports. 2017;**7**:17235

[44] Camara E, Albiol J, Ferrer P. Droplet digital PCR-aided screening and characterization of Pichia pastoris multiple gene copy strains. Biotechnology and Bioengineering. 2016;**113**:1542-1551

[45] Sugita S, Takase H, Nakano S. Practical use of multiplex and broad-range PCR in ophthalmology. Japanese Journal of Ophthalmology. 2021;**65**:155-168

[46] Shimizu T, Fujita T, Fujii H. Oligoribonucleotide interference-PCR:

Principles and applications. Biological Methods Protocol. 2022;7:bpac010

[47] Ku JL, Jeon YK, Park JG. Methylation-specific PCR. Methods in Molecular Biology. 2011;**791**:23-32

[48] Aono Y, Imai J, Tominaga K, Orita S, Sato A, Igarashi H. Rapid, sensitive, specific, and quantitative detection of human T-cell leukemia virus type 1 sequence in peripheral blood mononuclear cells by an improved polymerase chain reaction method with nested primers. Virus Genes. 1992;**6**:159-171

[49] Hudu SA, Alshrari AS, Syahida A, Sekawi Z. Cell culture, technology: Enhancing the culture of diagnosing human diseases. Journal of Clinical and Diagnostic Research. 2016;**10**:DE01-5

[50] McAdam AJ, Riley AM. Developments in tissue culture detection of respiratory viruses. Clinics in Laboratory Medicine. 2009;**29**:623-634

[51] Meyer B, Drosten C, Muller MA. Serological assays for emerging coronaviruses: Challenges and pitfalls. Virus Research. 2014;**194**:175-183

[52] Deng AP, Zhang YH, Sun LM, Zeng HR, Li W, Ke CW, et al. Survey on the recessive infection of pathogen to hand-foot-mouth disease among healthy adults and children in Guangdong province. Zhonghua Liu Xing Bing Xue Za Zhi. 2012;**33**:189-191

[53] Mahmmod YS, Toft N, Katholm J, Gronbaek C, Klaas IC. Estimation of test characteristics of real-time PCR and bacterial culture for diagnosis of subclinical intramammary infections with *Streptococcus agalactiae* in Danish dairy cattle in 2012 using latent class analysis. Preventive Veterinary Medicine. 2013;**109**:264-270

[54] Han JY, Han SB. Febrile seizures and respiratory viruses determined by multiplex polymerase chain reaction test and clinical diagnosis. Children (Basel). 2020;**2020**:7

[55] Le Calvez T, Trouilhe MC, Humeau P, Moletta-Denat M, Frere J, Hechard Y. Detection of free-living amoebae by using multiplex quantitative PCR. Molecular and Cellular Probes. 2012;**26**:116-120

[56] Zhong Q, Bhattacharya S, Kotsopoulos S, Olson J, Taly V, Griffiths AD, et al. Multiplex digital PCR: Breaking the one target per color barrier of quantitative PCR. Lab on a Chip. 2011;**11**:2167-2174

[57] Clifford G, Franceschi S, Diaz M, Munoz N, Villa LL. HPV type-distribution in women with and without cervical neoplastic diseases. Vaccine. 2006;**24**(Suppl 3):S3/26-34

[58] Carvalho Nde O, del Castillo DM, Perone C, Januario JN, Melo VH, Brasileiro FG. Comparison of HPV genotyping by type-specific PCR and sequencing. Memórias do Instituto Oswaldo Cruz. 2010;**105**:73-78

[59] Benhamou Y, Katlama C, Lunel F, Coutellier A, Dohin E, Hamm N, et al. Effects of lamivudine on replication of hepatitis B virus in HIV-infected men. Annals of Internal Medicine. 1996;**125**:705-712

[60] Paterson DL, Bonomo RA. Extended-spectrum beta-lactamases: A clinical update. Clinical Microbiology Reviews. 2005;**18**:657-686

[61] Ramirez MS, Tolmasky ME. Aminoglycoside modifying enzymes. Drug Resistance Updates. 2010;**13**:151-171

[62] Jacoby GA. AmpC beta-lactamases. Clinical Microbiology Reviews. 2009;**22**:161-182

[63] Saadatian Farivar A, Nowroozi J, Eslami G, Sabokbar A. RAPD PCR profile, antibiotic resistance, prevalence of armA gene, and detection of KPC enzyme in Klebsiella pneumoniae isolates. Canadian Journal of Infectious Diseases Medical Microbiology. 2018;**2018**:6183162

[64] Derakhshan S, Najar Peerayeh S, Bakhshi B. Association between presence of virulence genes and antibiotic resistance in clinical Klebsiella Pneumoniae isolates. Laboratoriums Medizin. 2016;**47**:306-311

[65] Pholwat S, Stroup S, Foongladda S, Houpt E. Digital PCR to detect and quantify heteroresistance in drug resistant mycobacterium tuberculosis. PLoS One. 2013;**8**:e57238

[66] Taylor SC, Carbonneau J, Shelton DN, Boivin G. Optimization of Droplet Digital PCR from RNA and DNA extracts with direct comparison to RT-qPCR: Clinical implications for quantification of Oseltamivir-resistant subpopulations. Journal of Virological Methods. 2015;**224**:58-66

[67] Mukaide M, Sugiyama M, Korenaga M, Murata K, Kanto T, Masaki N, et al. High-throughput and sensitive next-generation droplet digital PCR assay for the quantitation of the hepatitis C virus mutation at core amino acid 70. Journal of Virological Methods. 2014;**207**:169-177

[68] Tahamtan A, Ardebili A. Real-time RT-PCR in COVID-19 detection: Issues affecting the results. Expert Review of Molecular Diagnostics. 2020;**20**:453-454

[69] Antonarakis SE, Skotko BG, Rafii MS, Strydom A, Pape SE, Bianchi DW, et al. Down syndrome. Natural Review in Diseases Primers. 2020;**6**:9

[70] Alldred SK, Takwoingi Y, Guo B, Pennant M, Deeks JJ, Neilson JP, et al. First trimester ultrasound tests alone or in combination with first trimester serum tests for Down's syndrome screening. Cochrane Database of Systematic Reviews. 2017;**3**:CD012600

[71] Muncie HL Jr, Campbell J. Alpha and beta thalassemia. American Family Physician. 2009;**80**:339-344

[72] Lo YM, Chiu RW. Genomic analysis of fetal nucleic acids in maternal blood. Annual Review of Genomics and Human Genetics. 2012;**13**:285-306

[73] D'Aversa E, Breveglieri G, Boutou E, Balassopoulou A, Voskaridou E, Pellegatti P, et al. Droplet digital PCR for non-invasive prenatal detection of Fetal single-gene point mutations in maternal plasma. International Journal of Molecular Sciences. 2022;**23**:2819

[74] Constantinou CG, Karitzi E, Byrou S, Stephanou C, Michailidou K, Makariou C, et al. Optimized droplet digital PCR assay on cell-free DNA samples for non-invasive prenatal diagnosis: Application to Beta-thalassemia. Clinical Chemistry. 2022;**68**:1053-1063

[75] Chang MY, Ahn S, Kim MY, Han JH, Park HR, Seo HK, et al. One-step noninvasive prenatal testing (NIPT) for autosomal recessive homozygous point mutations using digital PCR. Scientific Reports. 2018;**8**:2877

[76] Guissart C, Tran Mau Them F, Debant V, Viart V, Dubucs C, Pritchard V, et al. A broad test based on fluorescent-multiplex PCR for noninvasive prenatal

diagnosis of cystic fibrosis. Fetal Diagnosis and Therapy. 2019;**45**:403-412

[77] SawakwongpraK,TangmansakulchaiK, Ngonsawan W, Promwan S, Chanchamroen S, Quangkananurug W, et al. Droplet-based digital PCR for non-invasive prenatal genetic diagnosis of alpha and beta-thalassemia. Biomedical Reports. 2021;**15**:82

[78] Han L. Genetic screening techniques and diseases for neonatal genetic diseases. Zhejiang Da Xue Xue Bao. Yi Xue Ban. 2021;**50**:429-435

[79] Zou Y, Dai QQ, Tao WJ, Wen XL, Feng DF, Deng H, et al. Suspension array-based deafness genetic screening in 53,033 Chinese newborns identifies high prevalence of 109 G>a in GJB2. International Journal of Pediatric Otorhinolaryngology. 2019;**126**:109630

[80] Garbuz MM, Ovchinnikova AA, Kumeiko VV. Design, Optimization and validation of the ARMS PCR protocol for the rapid diagnosis of Wilson's disease using a panel of 14 common mutations for the European population. Genes (Basel). 2022;**2022**:13

[81] Panagopoulos I, Andersen K, Rinvoll Johannsdottir IM, Micci F, Heim S. Novel MYCBP::EHD2 and RUNX1::ZNF780A fusion genes in T-cell acute lymphoblastic Leukemia. Cancer Genomics Proteomics. 2023;**20**:51-63

[82] Hur K, Cejas P, Feliu J, Moreno-Rubio J, Burgos E, Boland CR, et al. Hypomethylation of long interspersed nuclear element-1 (LINE-1) leads to activation of proto-oncogenes in human colorectal cancer metastasis. Gut. 2014;**63**:635-646

[83] Li XN, Wang ZJ, Ye CX, Zhao BC, Li ZL, Yang Y. RNA sequencing reveals the expression profiles of circRNA and indicates that circDDX17 acts as a tumor suppressor in colorectal cancer. Journal of Experimental & Clinical Cancer Research. 2018;**37**:325

[84] Tramm T, Mohammed H, Myhre S, Kyndi M, Alsner J, Børresen-Dale AL, et al. Development and validation of a gene profile predicting benefit of postmastectomy radiotherapy in patients with high-risk breast cancer: A study of gene expression in the DBCG82bc cohort. Clinical Cancer Research. 2014;**20**:5272-5280

[85] Tong Y, Ermel A, Tu W, Shew M, Brown DR. Association of HPV types 6, 11, 16, and 18 DNA detection and serological response in unvaccinated adolescent women. Journal of Medical Virology. 2013;**85**:1786-1793

[86] Wang H, Fu C, Du J, Wang H, He R, Yin X, et al. Enhanced histone H3 acetylation of the PD-L1 promoter via the COP1/c-Jun/HDAC3 axis is required for PD-L1 expression in drug-resistant cancer cells. Journal of Experimental & Clinical Cancer Research. 2020;**39**:29

[87] Wu L, Wang W, Zhang J, Wu X, Chen Y, Gu X, et al. Detection of five types of HPV genotypes causing Anogenital warts (Condyloma Acuminatum) using PCR-Tm analysis technology. Frontiers in Microbiology. 2022;**13**:857410

[88] Soverini S, Bassan R, Lion T. Treatment and monitoring of Philadelphia chromosome-positive leukemia patients: Recent advances and remaining challenges. Journal of Hematology & Oncology. 2019;**12**:39

[89] Fang Y, Yan D, Wang L, Zhang J, He Q. Circulating microRNAs (miR-16, miR-22, miR-122) expression and early diagnosis of hepatocellular carcinoma.

Journal of Clinical Laboratory Analysis. 2022;**36**:e24541

[90] Jin X, Gao J, Zheng R, Yu M, Ren Y, Yan T, et al. Antagonizing circRNA_002581-miR-122-CPEB1 axis alleviates NASH through restoring PTEN-AMPK-mTOR pathway regulated autophagy. Cell Death & Disease. 2020;**11**:123

[91] Shi J, Chen P, Sun J, Song Y, Ma B, Gao P, et al. MicroRNA-1258: An invasion and metastasis regulator that targets heparanase in gastric cancer. Oncology Letters. 2017;**13**:3739-3745

[92] Zhang Z, Li J, Huang Y, Peng W, Qian W, Gu J, et al. Upregulated miR-1258 regulates cell cycle and inhibits cell proliferation by directly targeting E2F8 in CRC. Cell Proliferation. 2018;**51**:e12505

[93] Zhang H, Jiang S, Guo L, Li X. MicroRNA-1258, regulated by c-Myb, inhibits growth and epithelial-to-mesenchymal transition phenotype via targeting SP1 in oral squamous cell carcinoma. Journal of Cellular and Molecular Medicine. 2019;**23**:2813-2821

[94] Behrouzi R, Barr CE, Crosbie EJ. HE4 as a biomarker for endometrial cancer. Cancers (Basel). 2021;**2021**:13

[95] Bray F, Ferlay J, Soerjomataram I, Siegel RL, Torre LA, Jemal A. Global cancer statistics 2018: GLOBOCAN estimates of incidence and mortality worldwide for 36 cancers in 185 countries. CA: A Cancer Journal for Clinicians. 2018;**68**:394-424

[96] Arnold M, Sierra MS, Laversanne M, Soerjomataram I, Jemal A, Bray F. Global patterns and trends in colorectal cancer incidence and mortality. Gut. 2017;**66**:683-691

[97] Flemer B, Lynch DB, Brown JM, Jeffery IB, Ryan FJ, Claesson MJ, et al. Tumour-associated and non-tumour-associated microbiota in colorectal cancer. Gut. 2017;**66**:633-643

[98] Zheng L, Yang Q, Li C, Xu G, Yuan Q, Hou J, et al. Ubiquitin-specific peptidase 8 modulates cell proliferation and induces cell cycle arrest and apoptosis in breast Cancer by stabilizing Estrogen receptor alpha. Journal of Oncology. 2023;**2023**:8483325

[99] Olmedillas-López S, Lévano-Linares DC, Alexandre CLA, Vega-Clemente L, Sánchez EL, Villagrasa A, et al. Detection of KRAS G12D in colorectal cancer stool by droplet digital PCR. World Journal of Gastroenterology. 2017;**23**:7087-7097

[100] Zhang C, Wei G, Zhu X, Chen X, Ma X, Hu P, et al. Exosome-delivered circSTAU2 inhibits the progression of gastric Cancer by targeting the miR-589/CAPZA1 Axis. International Journal of Nanomedicine. 2023;**18**:127-142

[101] Lin WH, Xiao J, Ye ZY, Wei DL, Zhai XH, Xu RH, et al. Circulating tumor DNA methylation marker MYO1-G for diagnosis and monitoring of colorectal cancer. Clinical Epigenetics. 2021;**13**:232

[102] Alumkal JJ, Zhang Z, Humphreys EB, Bennett C, Mangold LA, Carducci MA, et al. Effect of DNA methylation on identification of aggressive prostate cancer. Urology. 2008;**72**:1234-1239

[103] Harbeck N, Nimmrich I, Hartmann A, Ross JS, Cufer T, Grützmann R, et al. Multicenter study using paraffin-embedded tumor tissue testing PITX2 DNA methylation as a marker for outcome prediction in tamoxifen-treated, node-negative breast

cancer patients. Journal of Clinical Oncology. 2008;**26**:5036-5042

[104] Zhang Q, Hu G, Yang Q, Dong R, Xie X, Ma D, et al. A multiplex methylation-specific PCR assay for the detection of early-stage ovarian cancer using cell-free serum DNA. Gynecologic Oncology. 2013;**130**:132-139

[105] Ponomaryova AA, Rykova EY, Cherdyntseva NV, Skvortsova TE, Dobrodeev AY, Zav'yalov AA, et al. Potentialities of aberrantly methylated circulating DNA for diagnostics and post-treatment follow-up of lung cancer patients. Lung Cancer. 2013;**81**:397-403

[106] Bell AS, Ranford-Cartwright LC. Real-time quantitative PCR in parasitology. Trends in Parasitology. 2002;**18**:337-342

[107] Almeida-Oliveira F, Leandro JGB, Ausina P, Sola-Penna M, Majerowicz D. Reference genes for quantitative PCR in the adipose tissue of mice with metabolic disease. Biomedicine & Pharmacotherapy. 2017;**88**:948-955

[108] Miranzadeh-Mahabadi H, Miranzadeh-Mahabadi H, Nikpour P, Emadi-Baygi M, Kelishadi R. Comparison of TaqMan real-time and tetra-primer ARMS PCR techniques for genotyping of Rs 8066560 variant in children and adolescents with metabolic syndrome. Advances in Clinical and Experimental Medicine. 2015;**24**:951-955

[109] Song Y, Wu L, Li M, Xiong X, Fang Z, Zhou J, et al. Down-regulation of MicroRNA-592 in obesity contributes to hyperglycemia and insulin resistance. eBioMedicine. 2019;**42**:494-503

[110] Nepom GT, Erlich H. MHC class-II molecules and autoimmunity. Annual Review of Immunology. 1991;**9**:493-525

[111] Buck GE, Eid NS. Applications of the polymerase chain reaction (PCR) to the diagnosis of pediatric pulmonary disease. Pediatric Pulmonology. 1995;**20**:309-322

[112] Capon F, Melchionda S, Gennarelli M, Lo Cicero S, Giacanelli M, Novelli G, et al. A tool for the molecular analysis of an early lethal disease: Slide-PCR in spinal muscular atrophy patients. Molecular and Cellular Probes. 1993;7:221-226